SpringerBriefs in Fire

Series editor

James A. Milke, University of Maryland, College Park, USA

More information about this series at http://www.springer.com/series/10476

Francine Amon · Jonatan Gehandler
Selim Stahl · Mai Tomida
Brian Meacham

Development of an Environmental and Economic Assessment Tool (Enveco Tool) for Fire Events

 Springer

Francine Amon
SP Technical Research Institute of Sweden
Borås
Sweden

Mai Tomida
Worcester Polytechnic Institute
Worcester, MA
USA

Jonatan Gehandler
SP Technical Research Institute of Sweden
Borås
Sweden

Brian Meacham
Fire Protection Engineering
Worcester Polytechnic Institute
Worcester, MA
USA

Selim Stahl
SP Technical Research Institute of Sweden
Borås
Sweden

ISSN 2193-6595   ISSN 2193-6609 (electronic)
SpringerBriefs in Fire
ISBN 978-1-4939-6558-8   ISBN 978-1-4939-6559-5 (eBook)
DOI 10.1007/978-1-4939-6559-5

Library of Congress Control Number: 2016946932

Printed on acid-free paper

This Springer imprint is published by Springer Nature
The registered company is Springer Science+Business Media LLC New York

Foreword

In broad terms, the impact of fire on a community is usually measured in terms of the number of fires, human casualties, and property damage. There are, however, more subtle impacts of fire that are not so easily estimated but contribute to the measure of overall performance of the fire service in protecting a community. A simple method of estimating two of these issues, environmental impact and economic impact, is proposed to help fire departments communicate the value of their services to the communities they protect.

While environmental and economic impact assessment methodologies exist as separate systems, they generally require a high level of knowledge that is outside the scope of most fire departments. A relatively simple methodology for estimating the environmental and economic impact of fires will help communities understand the degree to which fire department activities can benefit a community's environmental and economic well-being.

The foundation undertook this study to investigate the feasibility of developing a tool that enables fire departments to estimate the value of their services to a community in terms of environmental and financial impact. This book provides a summary of this effort, which resulted in the development of a prototype tool for fire department use.

The Fire Protection Research Foundation expresses gratitude to the authors Francine Amon, Jonatan Gehandler, and Selim Stahl, who are with SP Technical Research Institute of Sweden located in Borås and Göteborg, Sweden and Mai Tomida and Brian Meacham, who are with Worcester Polytechnic Institute in Worcester, MA, USA. The Research Foundation appreciates the guidance provided by the Project Technical Panelists and all others that contributed to this research effort. Special thanks are expressed to the National Fire Protection Association (NFPA) for providing the project funding.

The content, opinions, and conclusions contained in this book are solely those of the authors and do not necessarily represent the views of the Fire Protection Research Foundation, NFPA, Technical Panel, or sponsors. The foundation makes no guaranty or warranty as to the accuracy or completeness of any information published herein.

Borås, Sweden   Francine Amon
Borås, Sweden   Jonatan Gehandler
Borås, Sweden   Selim Stahl
Worcester, USA   Mai Tomida
Worcester, USA   Brian Meacham

About the Fire Protection Research Foundation

The Fire Protection Research Foundation plans, manages, and communicates research on a broad range of fire safety issues in collaboration with scientists and laboratories around the world. The foundation is an affiliate of NFPA.

About the National Fire Protection Association (NFPA)

Founded in 1896, NFPA is a global, nonprofit organization devoted to eliminating death, injury, property and economic loss due to fire, electrical, and related hazards. The association delivers information and knowledge through more than 300 consensus codes and standards, research, training, education, outreach and advocacy; and by partnering with others who share an interest in furthering the NFPA mission.

All NFPA codes and standards can be viewed online for free.

NFPA's membership totals more than 65,000 individuals around the world.

Acknowledgments

The authors would like to acknowledge the following individuals who reviewed this book.

Project Technical Panel

Dave Butry, NIST
Jeff Case, Deputy Chief, Phoenix (Arizona) Fire
Janice Coen, National Center for Atmospheric Research
Shawn Davis, Chevron
Pravin Gandhi, UL
Chris Gallo, EPA
Kenan Ozekin, Water Research Foundation
Ned Pettus, Retired Fire Chief, Columbus Ohio
Debbie Smith, BRE
Chris Wieczorek, FM Global
Dave Waterhouse, Battallion Chief, Montreal Fire Department
Larry Williams, Fire Chief Dothan, Alabama
Meghan Housewright, NFPA
Christina Holcroft, NFPA

Project Sponsor

National Fire Protection Association

Contents

Chapter 1
Introduction and Background

In the US many aspects of the impacts of fire are calculated on a national level at regular intervals by the National Fire Protection Association (NFPA) and others [1–4] based largely on information consolidated in the National Fire Incident Reporting System (NFIRS) database and on annual surveys of fire departments. On a local level, the impact of fire on a community is usually measured in terms of the number of fires, human casualties, and property damage. There are, however, subtle impacts of fire that are not as easily estimated but contribute to the measure of overall performance of the fire service in protecting a community. A tool has been developed that provides a consistent methodology for assessing the performance of fire departments with respect to two of these impacts: environmental and economic. Rather than focus on costs and losses, this environmental and economic (Enveco) assessment tool expresses these impacts in terms of savings with respect to emissions to the environment and economic value.

Concern for the health of the natural environment is growing as human population grows and as new levels of contamination of scarce resources are revealed. Most fires in the built environment contribute to contamination of air and possibly also to surface water, groundwater, sediment, and soil [5–7]. Firefighting operations can also impact these same receptors [8]. Fire debris and ash often contain many harmful constituents, depending on the fuel and burning conditions of the fire [9]. From an economic perspective, the direct and indirect impacts of fire to a community can be devastating [10], but are not usually reported *at a local scale* beyond an account of the human deaths and injuries and the amount of property destroyed or damaged.[1] An estimate of the economic consequences of specific fire incidents that includes not only the impact on humans and property, but also such factors as fire protection, insurance, rebuilding of the structure(s), temporary housing, ability

---

[1]One can find many examples of this type of local fire damage reporting online, for instance: http://lacrossetribune.com/jacksoncochronicle/news/local/fire-damages-black-river-falls-home/article_9a60f130-4e72-5055-a837-1829686e2e88.html.

© Fire Protection Research Foundation 2016
F. Amon et al., *Development of an Environmental and Economic Assessment Tool (Enveco Tool) for Fire Events*, SpringerBriefs in Fire, DOI 10.1007/978-1-4939-6559-5_1

to continue working and doing business, etc., as described by Hall [1] could be of benefit to the fire service and other interested parties.

Most people deal with economic issues of some type nearly every day and so they have a sense of the value of economic savings. Relatively few people are familiar enough with measures of environmental performance to appreciate their magnitude without using some sort of comparison. In this work a baseline will be used to compare the environmental and economic consequences of fire service activities against a fictitious world in which the fire service exists but does not respond. This baseline is created for comparison purposes only. It is assumed that the same mix of career and volunteer firefighters exists, whether or not they respond. Quantitative risk assessment (QRA) is used to predict the fire behavior and damage for the theoretical baseline case in which the fire service does not respond. Results from the QRA are used in both the environmental and economic impact calculations.

The work reported here is the first step toward the development of the Enveco tool; it is a feasibility study. If the results of this work appear promising, further work will be conducted in the future to expand and improve the capabilities of the tool. At this stage, two relatively simple case studies were chosen to demonstrate the feasibility of the tool. In both case studies water was used as the fire suppressant in response to a warehouse fire in which hazardous materials were not stored (although there were oxidizers in one of the warehouses); this decision was made by the authors and officers of the Fire Protection Research Foundation (FPRF) and the National Fire Protection Association (NFPA) in order to simplify the demonstration. The tool is capable of being expanded to include more complex functionality in the future. Ideas for future expansion are the subject of Chap. 8, Future Work.

Additional background information is provided in the appendices as follows:

- Appendix A—Existing Risk, Environmental, and Economic Assessment Models
- Appendix B—Historic Warehouse Fires as Potential Case Studies
- Appendix C—Statistical Decision Support for Possible Future Expansion of the Enveco Tool.

1.1 Selection Criteria for Enveco Tool Components

Most existing risk and environmental assessment methods, and many economic assessment methods, require specialized knowledge that is outside the scope of most fire departments. The goal of this work is to develop a relatively easy-to-use methodology for estimating the environmental and economic impact of fires that will help communities understand the degree to which fire department activities influence a community's environmental and economic well-being.

The criteria for selecting which of the available methods of assessing environmental and economic impact to include in the Enveco tool were:

- Capability to handle fire
- Ease of use
- Availability of input data
- Quality of results
- Compatibility with other chosen methods and spreadsheet platform.

Discussion of the selection process is presented in the following sections. See Appendix B for additional background information about existing risk, environmental, and economic assessment models.

1.1.1   Environmental Assessment

Environmental and economic impact assessment methodologies already exist as both individual and combined systems. On the environmental side there is life cycle assessment (LCA), which is a methodology that can be used to evaluate the potential environmental impacts of a product, process, or activity. LCA is a comprehensive method for assessing impacts across the full life cycle of a product or system, from materials acquisition through manufacturing, use, and end of life. A formal procedure for conducting an LCA has been standardized by the International Organization for Standardization (ISO) in ISO 14040 and ISO 14044 [11, 12]. Other non-standardized techniques based on LCA have been developed to account for the effects of fire as an accidental end of life, which are not included in the ISO standards. One such method is the Fire-LCA model developed by SP in Sweden[2] [13-16] and later applied by others [17]; this method will be described in detail in a later section. Another method is the Dynamic LCAFire developed by Chettouh et al. [18]. The Dynamic LCAFire model includes a dispersion model for incorporating fire plumes into the LCA impact assessment calculations. In general, LCA-based environmental impact methods can be used to assess a wide range of environmental impact categories, for example: global warning, eutrophication, resource depletion, ecotoxicity of soil and water bodies, etc., depending on which impact assessment method is considered important for the goals of the LCA.

There are also methods such as carbon and water footprints that deal with a single type of impact. The procedure for conducting carbon and water footprints has been standardized in ISO 10467 [19] and ISO 14046 [20], respectively. There are numerous carbon and water footprint calculators available online that allow anyone with internet access to estimate their footprints. The US Department of Energy (DOE) has also developed calculation tools to estimate the carbon footprint and

---

[2]At SP Sveriges Tekniska Forskningsinstitut AB in Borås, Sweden.

energy index of a range of structures by comparison with entries in their Buildings Energy Data Book database [21].

Since fire effluents can have a negative impact on the quality of air, soil, and water an LCA-based environmental assessment method was chosen so that several types of impacts could be included in the assessment. The TRACI impact assessment method [22] was chosen because it was developed by the U.S. Environmental Protection Agency (EPA) for use within the United States, it is also the environmental assessment method used in an LCA model that evaluates structures (Athena, [23]) and because the TRACI method is also available in a traditional LCA model (SimaPro, [24]) that can be used for non-structure products such as the warehouse contents. The effects of fire were incorporated into the assessment using the Fire-LCA methodology developed by SP [14], which is well-documented and draws on readily available expertise. The procedure for estimating the environmental impact of a warehouse fire is described in detail in Sect. 3.3.

1.1.2   Economic Assessment

There are many available tools to assess the cost-effectiveness of alternative scenarios or configurations, including Life Cycle Costing (LCC), Input-Output economic modeling and Cost Benefit Analysis (CBA). LCC uses a life cycle thinking perspective on the process or scenario and is sometimes used for product purchasing strategies [25] but is not easily applied to incident analysis. Input-Output economic modeling requires knowledge of the economic network within which the effects of the fire reside. The connections between the business that suffered the fire and its suppliers, customers and employees are extensive and difficult to quantify. CBA is often used to estimate the strengths and weaknesses of alternative scenarios and is frequently combined with risk modeling [26] and fire assessment [27] to determine whether risk reduction measures should be implemented. CBA is also more familiar and accessible to users that are not professional economists. For these reasons, CBA was chosen as the method by which economic impacts are assessed in the Enveco tool.

1.1.3   Risk Assessment

The QRA methodology is based on the recommended approach of Ingason in [28], which is based on the work of Heskestad [29, 30] and Mudan and Croce [31] and is applied to warehouse fires. The QRA requires a fully developed compartment/warehouse fire and basic data about the fire conditions at the time fire spread is most likely to occur.

This method uses deterministic information from the known fire, such as the orientation and location of adjacent structures, the size and location of breaches in the burning warehouse roof, wind/weather conditions, etc. to inform a predictive model of fire spread potential for the unknown baseline case in which the fire service does not respond.

1.1.4   Combined Fire Impact Assessments

Previous research studies have investigated the combinations of environmental and economic impacts of structure fires to compare the differences made by fire protection systems. Researchers at Carnegie-Mellon University report that input-output economic models can be integrated with LCA calculations in assessment of product chains where cost is an important factor [32, 33]. In 2009 FM Global assessed carbon emissions with respect to the risk of fire and natural hazards in a range of structure types [34], although this study does not directly address economic impacts. BRE Global looked specifically at warehouse fires in their 2013 study in which a methodology was developed to estimate the carbon emissions of warehouse fires in England as a function of warehouse size. The study includes a CBA of warehouses with and without sprinkler systems [35].

The Enveco tool uses QRA methods to predict the fire spread to adjacent structures in the theoretical scenario in which the fire service does not respond to the fire. The result of the QRA is used by both the LCA and the CBA to estimate the savings in environmental and economic impact, respectively, as a consequence of fire service intervention.

Chapter 2
Scope

At this point, the Enveco tool has been developed only far enough to determine whether the usefulness of the tool outweighs the uncertainty of the results. A large number of assumptions were necessary to account for various aspects of the underlying model, which are discussed in detail in Chap. 3. The tool in this initial form is developed to be used on a case-by-case basis for warehouse fires in which water was the only suppression media used during the response. The Enveco tool analyzes fire service response to previously occurring warehouse fires in which a defensive strategy was used and makes a comparison with the predicted consequences of the same fire without fire service response. The tool does not predict response outcomes for future fires. Nor does it provide predictions about the amount of damage incurred as a function of response time or fire growth within the structure; there is no fire growth model involved aside from a QRA of the spread of the fire from the original structure to adjacent structures in cases without fire service intervention. Future versions of this tool may be capable of additional functionality, such as applicability to commercial and/or residential fires, including predictive results in terms of response time, ability to handle hazardous chemicals, etc. Ideas for potential future expansion are the subject of Chap. 8, Future Work.

The Enveco tool is designed to compare the environmental and economic burden of warehouse fires in cases when the fire service responds to the incident using water as the suppression agent with cases when the fire service does not respond to the same incidents. The burning warehouse is assumed to be located in an industrial park such that any adjacent structures that might be at risk from fire spread are similar to the burning warehouse with respect to size and occupancy. Consequently, if there is no threat of fire spread to adjacent structures and the fire service chooses to stand by without diminishing the fire damage to the burning warehouse, the environmental and economic impact of the response will be higher for the response case than for the baseline case. This result is due to deployment of equipment and personnel to the fire scene.

© Fire Protection Research Foundation 2016
F. Amon et al., *Development of an Environmental and Economic Assessment Tool (Enveco Tool) for Fire Events*, SpringerBriefs in Fire,
DOI 10.1007/978-1-4939-6559-5_2

The prototype Enveco tool does not currently consider these events/conditions:

- The effect or impact from explosions
- The release of toxic or radioactive material
- Types of structures other than one floor warehouse structures
- Civilian life safety (injuries or fatalities due to direct fire exposure or toxic fire effluents)
- Fire development as a function of time
- Fire service response time
- Averaged input from multiple incidents
- Human health effects due to environmental emissions.

Chapter 3
Approach

The Enveco tool is based on user input as well as underlying models obtained from the literature, technical reports, and software. This information provided by models is used to provide default values for situations when the user does not have direct knowledge of the input values.

The steps used in the development of the Enveco tool were:

- **Parameter identification**: Using the warehouse case studies as a guide, the parameters needed to perform a QRA, screening LCA, and CBA were determined. Existing and emerging risk, life cycle, and cost/benefit analysis methods were evaluated in terms of ease of use, type of parameters required, compatibility with other chosen assessment methods, and quality of results.
- **Inventory of structure and site types**: Incidents to which fire departments respond vary considerably, both in the type of structure and in the local conditions at a particular location. The types of structures/sites used by the tool are limited initially to a warehouse in an industrial park, however, information was collected on areas of future expansion.
- **Interface with fire service**: A project technical panel comprised of fire service and technical experts was formed early in the project to provide guidance and refine the needs of the fire service and the usability requirements of the tool.
- **Tool platform**: For this early phase in the development of the tool a spreadsheet was used as the platform. It incorporates a user input worksheet, an output worksheet for the results, and several calculation worksheets for the QRA, CBA, and LCA.
- **Case studies**: The Enveco tool was exercised on two well-documented historic warehouse fires. Both cases were large impact fires suppressed with water. The first case study will be treated as a generic warehouse fire having contents that conform to the NFPA 13 hazard occupancy class "Extra—Group 1", although it is based on a real warehouse fire. This warehouse had adjacent structures on all sides and is thus a good case to demonstrate the potential fire spread hazard handled by the QRA. The second case study was a warehouse fire that contained

© Fire Protection Research Foundation 2016
F. Amon et al., *Development of an Environmental and Economic Assessment Tool (Enveco Tool) for Fire Events*, SpringerBriefs in Fire, DOI 10.1007/978-1-4939-6559-5_3

household appliance parts that also fall into the "Extra—Group 1" hazard occupancy classification. In both case studies the warehouses were entirely destroyed by the fire.

- **Implementation and dissemination**: This work will be implemented initially through the end users in the reference group. It will also be submitted for publication in Fire Technology and presented at the 2016 NFPA Conference and Expo, the 2016 NFPA Symposium on Fire Protection for a Changing World, and the Sustainable Fire Engineering 2016 conference.

The three major components of the tool are QRA, LCA and CBA. These components are described in detail in the following sections. Figure 3.1 shows how the components of the tool fit together, along with their specific input requirements and output. A deeper discussion of the integration of these components into the comprehensive Enveco tool is provided in Chap. 4.

The analytical approach used in each of the three Enveco tool components (QRA, LCA, CBA) is described in the following sections.

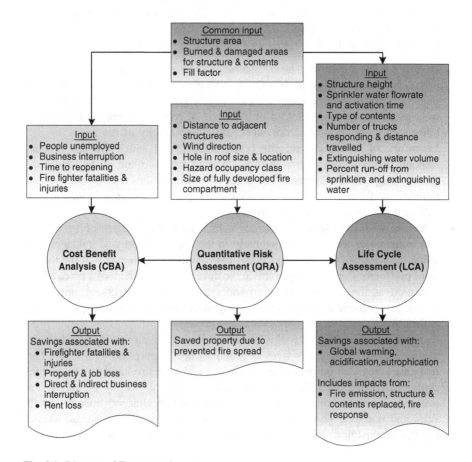

Fig. 3.1 Diagram of Enveco tool components

3.1 Quantitative Risk Assessment

The QRA estimates the risk of damage due to fire spread in the baseline case (without fire department intervention) in terms of likelihood, starting from information that is known about the actual fire (with fire department intervention). It estimates the fire spread to adjacent structures and the added fire damage that most likely would have occurred, e.g. if adjacent structures were not protected by the fire service. In other words, the QRA estimates the adjacent property that is saved as a result of fire service intervention. Response to warehouse fires is frequently defensive because the fire can grow rapidly in large open spaces having high fuel loads and the conditions can quickly become unsafe for firefighter entry [36].

It is assumed that the level of fire safety for buildings is the same whether or not the fire service responds to the fire. From a building regulation perspective there are two ways to address fire spread. Firstly, the generation and spread of fire and smoke within the construction should be limited. Secondly, the spread of fire to neighboring structures should be limited. In prescriptive terms this is translated to the use of fire rated material, compartmentation, fire walls, suppression systems and building space separation [37–39].

According to Campbell, some type of sprinkler system was present in 32 % of warehouse structure fires during the period from 2007 to 2011. In cases when wet pipe sprinkler systems were activated by the fire, they were effective 84 % of the time [40]. It can be inferred from fire investigation reports of the actual (fire service intervention) case whether an existing sprinkler system was activated and controlled the fire or not. Future versions of the Enveco tool may include analysis that can determine whether a quick intervention extinguished a fire that would have become a fully developed warehouse fire without intervention. For fire spread to adjacent buildings a probabilistic approach is developed based on a fully developed compartment/warehouse fire [40].

Basic theory about ignition due to radiation, the main mode of fire spread between buildings [28, 41], is given in the following subsections. For fire spread from warehouse structures, Ingason's recommended approach in [28], mainly based on the work of Heskestad [29, 30] and Mudan and Croce [31], will be explained in the next section. Other sources and models must be used for different structure types, e.g. [41, 42] study fire spread between offices and residential buildings.

3.1.1 Radiation and Ignition

Radiation is the main mode of heat transfer at high temperatures. The radiation, q, to a surface exposed to fire, S, is [43]:

$$q = \varepsilon(q_{inc} - \sigma T_s^4)$$

where ε is the emissivity of the surface (equal to the absorptivity), q_{inc} is the net incident radiation from all external sources, and σ is the Stefan-Boltzman constant and T_S is the surface temperature in Kelvin. q_{inc} can be equated to a weighted average of the surrounding temperatures, T_r.

$$q = \varepsilon\sigma(T_r^4 - T_s^4)$$

In general, the emissivity of most materials is between 0.8 and 0.9. The emissivity of a flame depends on the amount of soot particles and the flame thickness. For radiation between two infinite parallel planes, e.g. a facade fully involved in fire and an exposed facade having dimensions much larger than the separation distance, T_r becomes the temperature of the fire. For most cases though, neither the source nor the surface can be considered to be infinite planes. The exposed surface will see a cold surrounding and walls etc., and not just the fire surface area at a separation distance between them [43].

Most surfaces ignite by flames at surface temperatures between 250 °C (482 °F) and 450 °C (842 °F) and auto-ignite at around 500 °C (932 °F) without flame impingement. The time to ignition of thick homogenous materials is proportional to the thermal inertia, which is the product of specific heat capacity, density and conductivity. The time to ignition increases with increasing thermal inertia. Thus insulating materials ignite first, if they are combustible. For very thin solids the temperature may be assumed to be uniform. In this case the time to ignition is proportional to the product of density, specific heat capacity and material thickness. Flame spread can be seen as a continuous ignition process [43].

A fire is more likely to spread downwind. Wind can enhance the heat release rate (HRR) by providing oxygen and heat transfer by tilting the flames and smoke over or closer to other objects not yet on fire. For strong winds the cooling effect can start to dominate, although this is relative to the fire size [43]. In strong winds, warehouse fires are reported to have spread up to 125 m (410 ft), with flame lengths of at least 100 m (328 ft) [28].

A fire will spread to a nearby building if the cladding or roof is heated sufficiently to cause spontaneous ignition or piloted ignition in conjunction with embers [42]. Values reported for spontaneous ignition of timber varies from 28 kW/m^2 (2.6 kW/ft^2) to 33.5 kW/m^2 (3.1 kW/ft^2). The value for piloted ignition varies from 10 kW/m^2 (0.93 kW/ft^2) for long duration exposures to 18 kW/m^2 (1.7 kW/ft^2) for 30 min exposure [42]. Many regulations (e.g. [39]) use a lower value of 12.5 kW/m^2 (1.2 kW/ft^2) for piloted ignition of an adjoining façade, but according to Carlsson [41] piloted ignition due to exposure from full scale fires starts at higher radiation levels, 18 kW/m^2 (1.7 kW/ft^2) for unpainted wood and 26 kW/m^2 (2.4 kW/ft^2) for painted wood surfaces exposed to 15 min of radiation exposure. In general, Carlsson proposes that piloted ignition in full scale can occur at 15–18 kW/m^2 (1.4–1.7 kW/ft^2). According to [42], embers spread downwind in a 90° arc and at wind speeds higher than 20 km/h (12.4 mph), although it must be assumed that embers will also be present downwind from large warehouse flames at lower wind speeds. Based on [41] and [42] a limit of 18 kW/m^2 (1.7 kW/ft^2) is used for piloted ignition in the downwind direction and 33 kW/m^2 (3.1 kW/ft^2) for spontaneous ignition. It is assumed that the target structure for fire spread has the same roof height as that of the warehouse on fire.

3.1.2   Spread of Fire from Warehouse Fires

The main mode of fire spread between industrial buildings is through radiation. If no flashover fire is created, the risk of any type of break through the roof or windows is reduced and the risk of fire spread beyond the building of origin is limited. If the fire reaches complete flashover, considering the amount of fuel load, flames will burst through the roof unless the fire is extinguished. In this model it is assumed that the original burning warehouse is destroyed and that adjacent structures are similar warehouses. Future versions of the Enveco tool will include fire growth and estimate the importance of response time and staffing level in saving the original structure and contents.

The size and location of holes in the roof significantly affect the radiation and the risk of fire spread [28, 44]. According to [28] the recommended method for estimating the flame height and incident radiation is by the use of the following equations [29]. Note that all equations in this section are designed to be used with SI units.

$$H_f = -1.02D + 0.235\dot{Q}^{2/5}$$

where D is the diameter (in m) of the hole in the roof and $\dot{Q}$ is the total HRR (in kW). $\dot{Q}$ is estimated using the NFPA 13 classification of occupancies in [38]. Based on [28], the HRR per warehouse floor area (in m^2) is extrapolated from the NFPA 13 occupancy classification accordingly (Table 3.1).

Table 3.1 Definition of occupancy classes and their assumed HRR/m^2

Hazard occupancy class (NFPA 13)	Short description [38]	HRR (MW/m^2)
Light Hazard	Low HRR and fuel quantity	0.1–0.3
Ordinary Hazard (Group 1)	Moderate HRR and fuel quantity, 2.4 m (8 ft) high walls	0.3–0.5
Ordinary Hazard (Group 2)	High fuel quantity and high HRR with 2.4 m (8 ft) high walls or moderate HRR and 3.6 m (12 ft) high walls	0.5–0.8
Extra Hazard (Group 1)	Very high HRR and fuel quantity, rapid fire spread, but little or no flammable liquid	0.8–1.3
Extra Hazard (Group 2)	Moderate or substantial amount of flammable or combustible liquids	1.3–2.5

The incident radiation $\dot{Q}''$ at distance L from the center of the flame is:

$$\dot{Q}'' = X\frac{\dot{Q}}{4\pi L^2}$$

where $X = \dot{Q}_{rad}/\dot{Q}$ is the fraction of total heat release that is radiated to the surroundings. The value of X is often quoted as either 0.3 or as being within the range of 0.2–0.4 in literature [14]. The fraction of radiation is lowered by black smoke, which can be expected for most warehouse fires. Therefore X is assumed to be between 0.2 and 0.3 in the probabilistic model.

The tilt angle of the flame in wind, given a wind velocity of u m/s can be calculated by [31]:

$$\cos\theta = \begin{cases} 1, & for\ u^* \leq 1 \\ \frac{1}{\sqrt{u^*}}, & for\ u^* > 1 \end{cases}$$

where

$$u^* = u/\left(\frac{g\dot{m}''D}{\rho}\right)^{1/3}$$

And

$$\dot{m}'' = \frac{\dot{Q}}{XAH_c}$$

where g is the standard gravity, $\dot{m}''$ is the mass burning rate, ρ is the air density, A is the hole area and H_c is the heat of combustion. The tilting of the flame was not implemented in the current spreadsheet model, but it could be added at a later stage. With larger opening(s) in the roof, the HRR and flame height increases until the opening is large enough for the fire to become similar to a mass fire. Then the air induced or entrained by combustion dilutes the fuel vapors below their ability to burn and the flame is divided into several smaller local flames, this happens when H_f/D becomes less than about 0.5. The QRA model is based on the worst case for each fire simulation; it is assumed that the hole will start from a small diameter and enlarge gradually or in steps rather than the case where the whole ceiling collapses simultaneously. Openings in the center of the roof result in a more efficient combustion than openings close to the walls [28, 44]. Since the correlation between the position of the hole and the flame length is unknown, this factor is not included in the current model.

A probabilistic approach can be used in the case that it is unknown whether the roof collapsed in the actual fire. A roof collapse occurs in 10 % of warehouse incidents [36], however this is for all fires; the figure will be much larger for fully developed warehouse fires. It can be expected that an unprotected roof starts to break locally after 12 min from ignition for typical warehouse fires that are allowed to develop [14]. Considering the large fire load resulting in very long fires, it is assumed that holes do develop for fully developed warehouse fires, including those with protected ceilings/roofs.

To summarize, the following input is required to estimate the risk for fire spread from a warehouse to an adjacent building:

- Fully developed fire for warehouse

  - Length and width of the compartment/warehouse on fire (constants).
  - Heat release rate per unit area (uniform distribution).
  - Upper and lower limit for the hole diameter (worst case for each fire simulation, i.e. a diameter from approximately 10 m up to half the side length, and position (north, west, south, east, anywhere) of hole in the roof (uniform distributions).

- Wind direction (north, west, south and east) (constant probability).
- Distances from adjacent structure (up to four: north, west, south and east) to burning warehouse (constant).
- Target material ignition limit:

  - Piloted ignition (in wind direction): 18 kW/m^2 (1.7 kW/ft^2) (constant).
  - Spontaneous ignition (in all other directions): 33 kW/m^2 (3.1 kW/ft^2) (constant).

The origin of the Cartesian coordinate system is defined in Fig. 3.2. The adjacent buildings are assumed to be placed parallel along the whole length of each side of the burning warehouse. The cardinal directions are shown in Fig. 3.2 only as reference for the discussions in this book, it is not necessary to align the coordinate system with true north.

The hole in the roof is assumed to be circular and, depending on its radius, it can be placed according to Fig. 3.3a. If the location of the hole is known, its approximate location should be indicated according to Fig. 3.3b. If the location of the fire origin inside the warehouse is known, that is also where the first hole in the roof is most likely to appear. If, for example, the hole was in the west side of the warehouse roof, possible positions and radii will be simulated by the QRA model along the west side. If the location of the hole or fire origin is unknown, it will be simulated to be equally likely to occur anywhere in the green rectangle in Fig. 3.3a.

Flame height, flame tilting and radiation to adjacent targets are calculated in a spreadsheet using Monte Carlo simulations. The model uses some constant values or deterministic parameters without uncertainty intervals, e.g. the distance to adjacent structures. It also uses distributions and probabilistic parameters, e.g. the diameter and position of the hole in the roof. In the Monte Carlo simulation 10,000 simulations are run. For each run, the deterministic parameters are assigned their constant value and each uncertain parameter is assigned a random value base on its probabilistic distribution. The uncertain parameters have an equally plausible uniform distribution between two limits.

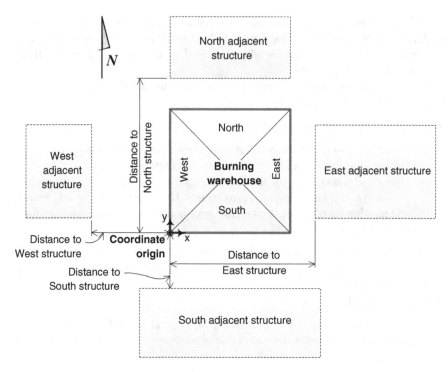

Fig. 3.2 Definition of point of origin and coordinate system (x, y) to determine distances to adjacent buildings

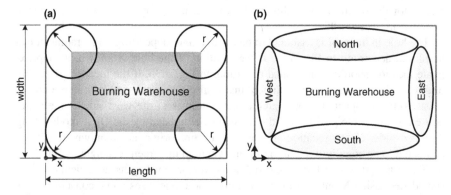

Fig. 3.3 a Possible placement of center of circular hole in the roof (*green rectangle*) depending on the radius of the hole and the dimensions of the warehouse. **b** Regions of the general location of the hole in the roof, "anywhere" can also be used if location is unknown

In the event that the fire spreads to an adjacent building, the probability that the adjacent building has a working sprinkler system is not part of the QRA model. Interior sprinkler systems are not effective against exterior fire threats.

3.2   Cost Benefit Analysis

In general, the process used to develop the CBA component of the Enveco tool generally followed guidance provided by Boardman [45]:

- **Define scope and comparison scenarios**: see Chap. 2, Scope, where both the scope of the project and the two scenarios are described.
- **Specify who will use the results**: the users were originally intended to be the fire service, however, other stakeholders such as community administrators, warehouse owners or managers, insurance companies, etc may also find value in using it.
- **Identify the indicators to be measured**: the indicators were chosen to be consistent with analyses conducted by others, and because they are of interest to the intended users of the Enveco tool. A description of the indicators is given below.
- **Assess the economic consequences**: this has been done for two case studies, see Chap. 6.
- **Use discount rate and cash flow analysis to calculate net present value (NPV)**: the US Consumer Price Index (CPI) was used to calculate NPV whenever source data was not current. All economic values are in 2016 United States dollars (USD).
- **Sensitivity analysis of the results**: see Chap. 5, Sensitivity and Uncertainty Analyses.
- **Recommend specific action**: see Chap. 7, Conclusions, and Chap. 8, Future Work.

3.2.1   Economic Indicators

The economic indicators used in the CBA are described in the following sections.

3.2.1.1   Firefighter Fatalities

It is a very difficult task to assign a monetary value to a human life. The Enveco tool uses a value of $9 million/life for fatalities, in accordance with Hall [1].

3.2.1.2   Firefighter Injuries

Firefighter injuries were estimated as 30 % of civilian injuries caused by fire, applied to an average cost of $16,600 per civilian injury in 1993 [1]. The NPV becomes $8257/injury in 2016 USD.

3.2.1.3 Replace Damaged Property

There is no accurate way of predicting the contents of the warehouse without prior knowledge. This information should be estimated from information provided by the warehouse owner or manager. Averaged data, weighted by number of fires per warehouse size category, from Table 15 in Campbell [40] was used to estimate the cost of replacing the damaged warehouse structure and contents. The NPV becomes $166.8/m^2$ in 2016 USD.

3.2.1.4 Fire Service Response

The cost of the response should include the costs of personnel and vehicles, allocated to the time spent responding to the incident. These costs depend heavily on the specific fire department. *Judgement of the user is necessary for this estimate.* A value of $600/truck/h was used, based on a sampling of cost recovery articles found on the internet.[1]

3.2.1.5 Job Disruption

This indicator includes lost wages and benefits for an unemployment duration of 29.1 weeks according to Bureau of Labor Statistics.[2] The annual salary range is $19,000–$39,000 per year for warehouse workers in the US. Using $29,000 as the midpoint, this result is $16,229/person. The number of people unemployed is estimated based on warehouse area. BRE Global estimates $62 m^2$ per person but uses a conservative value of $30 m^2$ for their analysis in [35]. Both values include effects beyond the burned warehouse, e.g., on suppliers, customers, transport. The Enveco tool uses a middle value of $50 m^2$.

3.2.1.6 Direct Business Interruption

Business interruption is averaged to 17 % of property damage using data from Table 9 in [35]. This estimation is based on highly uncertain data. There may not be a valid correlation between property damage and business interruption. Also the length of time of the interruption is not considered. The best source of this information is the warehouse owner or manager.

---

[1]For example: http://www.telegraph.co.uk/news/uknews/9261609/Calling-out-fire-brigade-will-now-cost-400-per-hour.html.

[2]See http://www.bls.gov/news.release/empsit.t12.htm.

3.2.1.7 Indirect Business Interruption

This indicator includes the effects of the fire on warehouse suppliers and customers. This is estimated as 10 % of direct business interruption in accordance with [1].

3.2.1.8 Rent Reduction

Rent reduction applies to the warehouse owner's loss of income due to the inability to rent warehouse space because of the fire. An average of random samples of warehouse rental costs[3] in the US resulted in a value of $0.4/sf.

3.3 Life Cycle Assessment

As applied to this work, LCA is used as a tool for evaluating the potential environmental impacts of a product, process, or activity [46], although there are other variations of LCA that focus on social, ecological, and other types of impacts. In general, LCA is a method of assessing impacts across the full life cycle of a product or system, from materials acquisition through manufacturing, use, and end of life. Depending on the application it is possible to examine the impact of only part of the life cycle, for example from cradle to gate, where the gate is some point in the life of the system being studied beyond which the life cycle has no further bearing. As depicted in Fig. 3.4, a standard LCA study is structured to have four major components: Goal and Scope Definition, Inventory Analysis, Impact Assessment, and Interpretation of results. The development of an LCA is typically an iterative process in which each of these components is revised as new information from other components is acquired.

The life cycle phases of a product or a system are assessed with respect to their impact on the environment (both good and bad) within this structure. The life cycle phases depend on the product or system but, for products, generally follow this pattern:

- Production (includes materials and manufacturing processes),
- Use (includes energy requirements, maintenance, during service life), and
- End of life (includes landfill, incineration, recycling).

The product or system being assessed could be nearly anything, for example LCA can be applied to the production of a warehouse (all or just part of it), or it could be used by politicians to examine the environmental consequences of policies and regulations, or it could be applied to internal industrial systems to, for example, optimize waste streams within a manufacturing facility. In this work, LCA is

---

[3]See http://warehousespaces.com.

Fig. 3.4 Structure of an LCA study

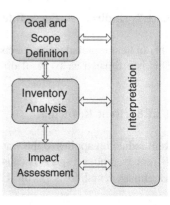

applied to the *production* of the warehouse and its contents, and also includes the *system* of fire service response and fire emissions to the environment.

Typically, LCA results are not expressed in monetary terms. The type of results depend on the purpose of the LCA but are usually reported in terms of environmental impact, such as climate change using CO_2 equivalents or acidification of soil and water using SO_2 equivalents. These results can stand alone, but are frequently used in comparisons between products or systems within the same study.

Accidental end of life by fire is not normally included in the ISO guidelines, so the application of LCA to a warehouse fire requires deviations from standard LCA practice in order to capture the environmental effects of the fire effluent and the activities of the fire service. The methodology will be described in the following sections using the same format as shown in Fig. 3.1.

3.3.1   Goal and Scope

The goal of this study is to assess and compare the environmental impact of a warehouse fire using two scenarios:

1. The fire service responds to the fire.
2. Fire service exists but does not respond.

In the first scenario a fire starts in a warehouse and the fire service responds, using water as the fire suppression media. It is assumed that this is a known incident for which information is available from reports, investigations, and/or interviews with involved parties such as first responders, owners or managers of the warehouse, and fire investigators.

In the second scenario a fire starts in the same warehouse under the same conditions. It is assumed that the fire continues to burn without intervention from the fire service and may spread to adjacent structures.

This study assesses the impact of replacing the damaged portions of the warehouse structure and contents, including adjacent structures and contents if the fire spreads. It also includes the impact of the fire effluent coming from the burned structures and contents and from firefighting operations. Finally, the impact of the existence of the fire service response (equipment, personnel and suppression media) is included in the first scenario, but not in the second scenario.

Replacement of the burning warehouse is modelled from cradle to gate. The gate is positioned at the end of the construction stage, meaning that the model includes extraction of raw materials from the earth, transport of the materials to refining facilities, additional processing of the materials in one or more facilities, transport of the finished materials to distribution centers (warehouses), transport to the construction site, and finally construction of the warehouse. It is assumed that a fire will occur in both scenarios, therefore the LCA model stops at the end of the construction phase[4] and does not include the use or the non-accidental end of life phases because they would be the same in both scenarios and do not contribute to the comparison.

Replacement of all or part of the contents is modelled from cradle to gate as well. For the contents the gate is positioned at the point when the contents arrive at the warehouse, meaning that the model includes the same steps listed for the burning warehouse above starting at extraction of materials and continuing through each step until they arrive at the warehouse in this study.

The fire effluents are modelled using a combination of the EcoInvent 3.0 database and the results of fire tests.

The fire service exists in both scenarios but is activated to respond to the fire only in the first scenario, thus the environmental impact of the response is captured and contributes to the comparison of the two scenarios. The model includes the number of vehicles and distance travelled to/from the event and the amount of fire suppression media used (in this case water only).

3.3.2 Inventory Analysis

Quite a lot of information (inventory data) is needed in order to assess the environmental impact of a warehouse fire. This information may come from observations or reports generated by the fire service, fire investigation reports, or it can be gathered from insurance companies or from the owners' inventory and architectural plans. The quality of the LCA model depends heavily on the accuracy and completeness of the inventory data, which can be difficult to obtain.

There are three main types of impacts: those associated with the impact of replacing damaged structures and contents, those associated with the fire effluent, such as pollutants entering the environment via the fire plume, and those associated directly with the fire response.

---

[4]This phase in the life of a warehouse is analogous to the production phase discussed in the previous section.

3.3.2.1 Warehouse Structure Replacement

The warehouse inventory data includes the raw materials, transport, manufacturing, and construction-installation processes associated with construction of a 9290 m^2 (100,000 sf) warehouse using the metal building systems (MBS) technique. The foundation is a 202 mm (8 in) thick concrete slab and there is 101 mm of fiberglass batt insulation covered by 13 mm (½ in) gypsum board in the walls and roof. This construction type was chosen for this phase of the Enveco tool development because it is common among warehouses; in the future it will be possible to include other construction materials/techniques.

The Enveco tool user inputs the total area of the warehouse footprint and the percent of warehouse structure (in terms of footprint area) that requires replacement due to damage from the fire. This input is used to scale the warehouse structure inventory data from the modelled warehouse up or down to the size of the warehouse being assessed. It is assumed that the warehouse is single story with a wall height of 9 m (30 ft). The storage racks are not included in the warehouse structure model.

3.3.2.2 Contents Replacement

The contents requiring replacement due to damage from the fire are calculated using user input of the warehouse footprint area and the percent of fire damaged contents in terms of footprint area. Additional user input is needed to estimate the amount and type of contents, including the percent of warehouse area that is filled with contents (the fill factor), the storage rack height, and the mix of materials.

It is not possible to predict the type of contents that might be stored in a warehouse; therefore some materials have been selected that are generally representative of components of a range of products. This is not a perfect solution because there can be large differences in impact between similar products and because the material categories don't represent a finished product consisting of many parts and combinations of materials. The Enveco tool probably underestimates the environmental impacts associated with replacing the warehouse contents because the tool only includes generic or surrogate items and not the production of specific items. For example, the impact of producing a gardening tool that is made of metal and wood and has plastic and paper packaging will not include the processing steps to heat and extrude the plastic and metal parts to make the finished tool, nor will it include the waste streams coming from the production processes, nor will it include the cutting and assembly of the plastic and paper packaging.

The selected material categories are:

- Plastic (*Polyethylene low-density film*; ex: packaging, products, film wrapping)
- Paper (*Kraft paper*; ex: packaging, products, cardboard)
- Wood (*Pine*; ex: products, pallets)
- Textile (*Woven cotton*; ex: any kind of textile except plastic-based)

- Soft furniture (*Polyurethane flexible foam*; ex: anything with foam cushions)
- Chemicals (*Alkyd paint*; ex: used to represent any "chemical")
- Dry food (*Flour*; ex: human or animal food)
- Electronics (*Generic component*; ex: consumer electronics such as TVs or computers, also components).

These material categories can be expanded and refined in the future. The user inputs an estimate of the percentage of damaged contents represented by each category of material. The volume is calculated using the warehouse footprint area, percent of damaged contents, rack storage height, and the fill factor. Densities of each material are then used to calculate the total weight of each material category.

The Enveco tool does not have its own dedicated database so the results from the EcoInvent 3.0 database were obtained using SimaPro software and were exported into the Enveco tool spreadsheet. These results are linear and can be scaled according to the weights calculated from the user input described above.

3.3.2.3   Fire Effluents

Characterizing the fire effluent is a challenge because it is dependent on the fuels involved, the ventilation conditions, and the suppression activities. For this reason it is difficult to predict the composition, concentration, and fate of pollutants coming from the fire. Measurements made by Blomqvist et al., in which fire effluent to air from fire tests of furnished rooms [47] and fire effluent to water from various non-industrial structural fire test results [48] were are used to represent the fire effluent for both the warehouse structure and the contents. Based on the previous work, the yields used in this study are listed in Table 3.2.

Table 3.2 Pollutants found in fire effluents to air and fire water run-off

Pollutant	Air [47] (kg/m^2 of burning structure)	Water^a [48] (mg/l of fire water run-off)
Ammonia (NH$_3$)	0.026	n/a
Carbon dioxide (CO$_2$)	26.8	n/a
Dioxins (total)	4.2×10^{-9}	n/a
Furans (total)	8.0×10^{-9}	n/a
Hydrogen chloride (HCl)	0.0406	n/a
Nitrogen oxides (NO)	0.0	n/a
Sulfur dioxide (SO$_2$)	0.1738	n/a
Polyaromatic hydrocarbons (PAH)	0.0119	0.04
Volatile organic hydrocarbons (VOC)	0.0256	0.78

^{a}*Note* These values are averaged from 4 fire tests of furnished non-industrial structures. The VOC category includes semi-VOC pollutants in the fire water run-off. A wide range of heavy metals were also found in [48] but were not presented in absolute values, thus it was not possible to determine their concentrations

The amount of pollutants emitted to air is based on the percent of the warehouse footprint area in which the fire consumed the warehouse and contents. The amount of PAH and VOC in sprinkler water run-off to the environment (in mg) is calculated from user input of the sprinkler water discharge rate, length of operation time, and the estimated percent of sprinkler water run-off that escaped into the environment.

Ideally, the fire effluents used in the Enveco tool would be directly related to the construction type of the warehouse structure and the actual contents. As more information becomes available from fire tests and models the fidelity of the fire effluent will improve.

3.3.2.4 Fire Department Response

In scenario 1 the fire service responds to the warehouse fire; in scenario 2 the fire service exists but does not respond to the fire. It is understood that scenario 2 would probably not ever happen in real life but it is necessary to create a baseline for comparison of the environmental impact of the response to the fire. This means that the environmental impact associated with the existence of the fire department is shared in both scenarios, but the response to the warehouse fire is only a part of scenario 1. Therefore, the impacts associated with parking lots, training facilities, maintenance, energy use, waste streams, structures and contents of the fire station(s) are the same for both scenarios and can drop out of the analysis. The actual equipment that is used to respond to the fire (quantity and type of vehicle and distance travelled) is assigned to scenario 1 but isn't present in scenario 2, so it will appear in the comparison. The production and maintenance of the trucks is not included in the LCA model because these aspects of the truck life cycle would be the same for both scenarios.

The amount of PAH and VOC in fire water run-off to the environment (in mg) is calculated from user input of the estimated percent of total water used by the fire service that escaped into the environment.

For this early development of the Enveco tool all vehicles that respond to a warehouse fire are modelled as "Transport, freight, lorry 16–32 metric ton, EURO4". It is assumed that each truck carries 3.8 metric tons of water. It is assumed that 50 % of the trucks pump 3785 lpm (1000 gpm) for 75 % of the time they are at the incident. The ability to include a variety of responding vehicle types can be added to the tool in the future.

3.3.3 Impact Assessment

For this project the Athena Impact Estimator for Buildings v.5 was used to estimate the environmental impact of the warehouse structure and the fire station(s). This software uses the "Tool for Reduction and Assessment of Chemical and Other Environmental Impacts" (TRACI 2.1) impact assessment method. The SimaPro 8 software, again using the TRACI 2.1 impact assessment method, was used for the warehouse contents, fire effluents, and the fire engines. The impact categories available in the Traci 2.1 method are described in Table 3.3. Not all of the available

Table 3.3 Description of Traci 2.1 impact categories [49]

Impact Category	Units	Comments/description
Global warming	kg CO_2 equiv	Global warming is an average increase in the temperature of the atmosphere near the Earth's surface and in the troposphere, which can contribute to changes in global climate patterns. Emissions to air are weighted relative to the global warming effect of 1 kg of CO_2
Acidification	kg SO_2 equiv	Acidification is the increasing concentration of hydrogen ion (H+) within a local environment. Emissions to air and water are weighted relative to the acidification effect of 1 kg of SO_2
Eutrophication	kg N equiv	Eutrophication is the enrichment of an aquatic ecosystem with nutrients (nitrates, phosphates) that accelerate biological productivity (growth of algae and weeds) and an undesirable accumulation of algal biomass. Emissions to air and water are weighted relative to the eutrophication effect of 1 kg of nitrogen (N)
Ozone depletion	kg CFC-11 equiv	Ozone has effects on human health, crops, other plants, marine life, and human-built materials. Emissions to air are weighted relative to the ozone depletion effect of 1 kg of the refrigerant CFC-11 (also known as trichlorofluoromethane and R-11)
Smog	kg O_3 equiv	Ground level ozone is created by various chemical reactions, which occur between NOx and VOCs in sunlight. Ecological impacts include damage to various ecosystems and crops. Emissions to air are weighted relative to the smog production effect of 1 kg of O_3
Ecotoxicity	CTUe equiv	Freshwater ecotoxicity potentials of emissions to air, surface water, and soil are weighted according to the USEtox model, which is based on pollutant concentrations at which 50 % of a species population displays an effect (e.g. mortality) [50]
Carcinogenics[a]	CTUh	Cancerous effects on humans, based on USEtox model. Not used in the Enveco tool
Non carcinogenics[a]	CTUh	Non-cancerous effects on humans, based on USEtox model. Not used in the Enveco tool
Respiratory effects[a]	kg PM2.5 equiv	Small particles in ambient air have the ability to cause negative human health effects including respiratory illness and death. Not used in the Enveco tool
Fossil fuel depletion	MJ surplus	Fossil fuel depletion was not included as a separate impact category in the Enveco tool because many energy sources contribute to the energy calculations for structure and content replacement

The gray categories are not included in the Enveco tool
[a]Note that human toxicity is not included in this version of the Enveco tool, therefore impact categories that apply solely to human health are not used

impact categories are used in the Enveco tool for two reasons: first, it was decided early in the development of the tool that human health effects are outside the scope of the tool; and second, fossil fuel depletion was not included as a separate impact category because other energy sources are included in the total energy calculations.

3.3.4 Interpretation

The interpretation step in LCA involves analysis of the completeness and accuracy of the modelling process as well as analysis of the results. Conclusions and recommendations are made only after the model and results have been examined and the strengths and weaknesses identified. Sensitivity and uncertainty analyses are provided for all the components of the Enveco tool in Chap. 5. Analyses of the LCA model results for Case Study 1 and Case Study 2 are found in Sects. 6.1.3 and 6.2.3, respectively.

3.3.4.1 LCA Model Strengths

The primary strength of the LCA component of this tool is that non-environmental experts can use it to estimate the environmental impact of a limited number of warehouse fires, comparing scenarios in which the fire service intervenes with scenarios in which there is no fire service intervention. At the time of writing this book no references were found in the literature to indicate that this capability existed previously.

The model includes replacement of all or part of the warehouse structure and contents, fire effluent, and the fire response. It produces impact results in terms of global warming, acidification, eutrophication, ozone depletion, smog, ecotoxicity, and energy use. This breadth of scope is much wider than other environmental impact estimator tools that are accessible to non-LCA experts.

3.3.4.2 LCA Model Weaknesses

Trade-offs in model accuracy are necessary when simplifying a complicated assessment process such as LCA. By scientific and engineering standards, LCA has a relatively high level of uncertainty that can be exacerbated by simplifications and assumptions, thus making the results meaningless.

The most tenuous assumptions in the LCA model are those related to the warehouse contents and the fire effluent. The warehouse contents are modelled as materials, not whole products. There is a very large difference among the types of materials within each category. For example, low density polyethylene (LDPE) film is used to represent plastic. There are many forms of LDPE as well as many types of non-LDPE plastics, which may have additives that greatly affect their environmental impact.

It is possible that materials are stored in the warehouse rather than whole products; however, it is more likely that products comprised of several materials are stored in the warehouse. Excluding the production of whole products removes a major contributing phase of the product's life from the impact calculation.

The fire effluents are derived from fire tests of furnished non-industrial structures. There is no connection between the yields listed in Table 3.2 and the materials used in the warehouse structure or contents. It is unclear how to make this connection without requiring the user to provide an enormous amount of inventory data that would be difficult to obtain.

The geographical region assumed by the LCA model is Canada and the United States. It may be possible for future versions of the Enveco tool to increase the geographical resolution, thus causing the results to reflect regional differences more accurately.

Chapter 4
Integration and Implementation

The Enveco tool makes use of a simple spreadsheet platform. The user enters data in the "Input" worksheet that is divided into four areas: Risk of Fire Spread, Warehouse Description, Contents Description, and Fire Service Response. These input areas are discussed in more detail in the following sections. The white cells in the "Value" column allow the user to input data. Pop-up comments are used with some of the white cells to instruct the user, for example to input only positive numbers. Metric or British units can be selected in the "Unit" column for relevant input. Default values are given whenever possible and are based on referenced literature. The only cells the user can change are the white input cells and, in some cases, the units cells.

The results are shown on the "Output" worksheet for the three components of the Enveco tool. Complete sets of results are given for the baseline scenario in which the fire service does not respond, the scenario in which the fire service does respond, and the comparison of the two scenarios in which the savings due to fire service intervention are indicated. The full output worksheets for the two case studies are provided in Appendix A.

In addition to these two worksheets, there are five other worksheets used for calculations (one each for QRA, CBA, and LCA, one for content inventory data, and one for fire truck inventory data). These worksheets are available for viewing but are locked so that the user cannot accidentally change the calculations. There is one hidden worksheet that has lists of named cells that are used to create the drop-down menus for selecting units and directions in the input worksheet.

4.1 Risk of Fire Spread

There are two diagrams located to the right of the data input cells, one diagram shows how to lay out the coordinate system for determining distances to adjacent structures and the other diagram is an example of a "wind rose" obtained from the

© Fire Protection Research Foundation 2016
F. Amon et al., *Development of an Environmental and Economic Assessment Tool (Enveco Tool) for Fire Events*, SpringerBriefs in Fire, DOI 10.1007/978-1-4939-6559-5_4

National Oceanographic and Atmospheric Administration (NOAA). Users can go to the referenced NOAA website and generate a wind rose, which indicates wind direction and speed for the general location and date of the fire if this information is not available locally.

The user can enter distances to adjacent structures from the coordinate system origin in either meters or feet. An assumed northerly direction is used to make references within the tool to the direction of the wind, the locations of adjacent structures and a hole in the roof of the burning warehouse (if a hole in the roof exists).

The user also chooses a hazard occupancy class based on the mix of items stored in the warehouse, which is used to estimate the HRR/m^2 of the fire. The user can also indicate the size of the fully developed fire compartment if it was a different size than the burning warehouse footprint.

All of the above input, along with the warehouse footprint area and dimensions, is used to estimate the fire spread to adjacent structures. The results of this section of the tool are then used to calculate the environmental and economic impacts in the baseline scenario in which the fire service does not response to the warehouse fire.

4.2   Warehouse Description

The user inputs information about the warehouse in this part of the input worksheet. This information includes the dimensions of the burning warehouse, the storage rack height, the amount of water discharged by the sprinkler system if there was one, and the percent of the warehouse footprint area that was burned or sustained smoke/water damage. These data are used to calculate the environmental impact of the fire effluents and of replacing the damaged structure.

Additionally, the user inputs the number of people unemployed by the warehouse fire, the monthly losses associated with business interruption, and the amount of time the warehouse was not functional during reconstruction. These data are used in the CBA.

4.3   Contents Description

The user inputs the fill factor of the burning warehouse and the percent of warehouse footprint area that contained burned and damaged contents. The user also inputs estimates of the percent of materials that make up the contents stored in the warehouse. These data are used to calculate the environmental impact of the fire effluents and of replacing the damaged contents.

4.4   Fire Service Response

The user inputs the number of fire trucks that responded to the fire, the average distance driven to the warehouse (one-way), an estimate of the amount of water used and the percentage of this water that escaped into the environment. This information is used in the LCA model to estimate the environmental impact of the fire service response.

The user is also requested to enter the number of firefighter fatalities and injuries directly resulting from the warehouse fire. This information is used in the CBA model.

Chapter 5
Sensitivity and Uncertainty Analyses

The three components of the Enveco tool each have different methods of expressing their sensitivity to variations in input parameters and in estimation of the uncertainty of their results. In this section the individual sensitivity/uncertainty analyses are discussed first and then, in an overall sense, sensitivity to variations in the input parameters is examined.

5.1 QRA

The QRA calculations allow use of Monte Carlo simulations to determine the sensitivity and uncertainty of the results. The simulations are run 10,000 times for each condition (in this case 10,000 times for each direction from the burning warehouse).

The model includes the following user input parameters that are treated as constants:

- Distances from origin to threatened structures.
- Wind direction.
- Fire compartmentation.

The model includes the following user input parameters that are treated as stochastic variables with uncertainties modelled as uniform distributions:

- Location of the hole in the roof, e.g. south means the hole can be distributed along the south side (y_hole).
- Lower and upper diameter of the hole in the roof, e.g. uniform distribution between 15 and 30 m (r_roof).
- Hazard occupancy class. e.g. the interval defined by one of the five classes from Table 3.1 (D_HRR).

© Fire Protection Research Foundation 2016

F. Amon et al., *Development of an Environmental and Economic Assessment Tool (Enveco Tool) for Fire Events*, SpringerBriefs in Fire, DOI 10.1007/978-1-4939-6559-5_5

The model includes one non-input parameter (X, the ratio between radiated and total HRR that is modeled as a uniform stochastic parameter with an uncertain interval between 0.2 and 0.3). The spread in results and the sensitivity to the uncertainty in the input (tornado graphs) are shown in Figs. 5.1, 5.2, 5.3, 5.4, 5.5, 5.6, 5.7 and 5.8. The input was taken from Case Study 1 (see Chap. 6).

From the output of the incident radiation levels towards adjacent buildings (Q_N, Q_S, Q_E, Q_W) the risk of fire spread can be inferred using the limit of 18 kW/m^2 (1.7 kW/ft^2) in the wind direction (70 % towards east structure and 30 % towards south structure in Case Study 1) or 33 kW/m^2 (3.1 kW/ft^2) without

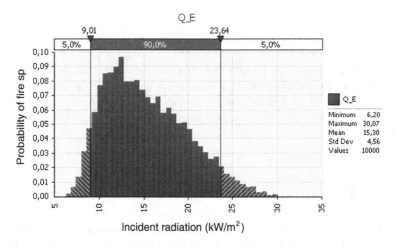

Fig. 5.1 Distribution of fire spread potential as a function of radiation level in the easterly direction from the burning warehouse. Graph was generated using @Risk software

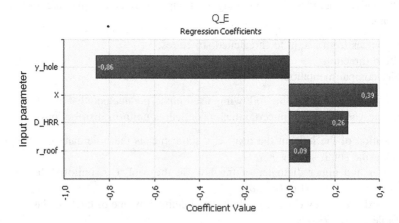

Fig. 5.2 Sensitivity of hole location (*y_hole*), *X*, hazard occupancy class (*D_HRR*), and hole size (*r_hole*) to the fire spread potential in the easterly direction. Graph was generated using @Risk software

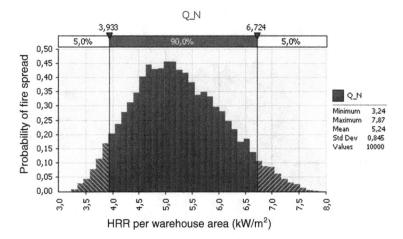

Fig. 5.3 Distribution of fire spread potential as a function of radiation level in the northerly direction from the burning warehouse. Graph was generated using @Risk software

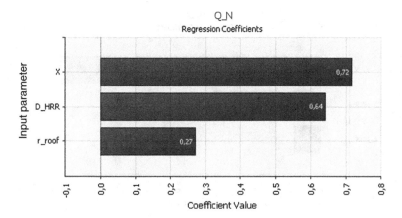

Fig. 5.4 Sensitivity of X, hazard occupancy class (D_HRR), and hole size (r_hole) to the fire spread potential in the northerly direction. Graph was generated using @Risk software

any help from the wind. This is a probabilistic distribution because not all of the input parameters are known with certainty.

Given the input from Case Study 1, the probability of fire spread results shown in Fig. 5.1 indicate that the lower limit of 18 kW/m^2 (1.7 kW/ft^2), which is relevant in this wind direction, presents a 36 % probability of fire spread (the area under the curve from 18 kW/m^2 and above). Without help from the wind, no ignition will occur since the area under the curve above 33 kW/m^2 is zero. As seen in Fig. 5.1, there is 90 % confidence that the mean incident radiation is between 9.01 and 23.64 kW/m^2.

The sensitivity of the QRA model input in the easterly direction is shown in Fig. 5.2. The location of the hole in the roof is by far the most sensitive input

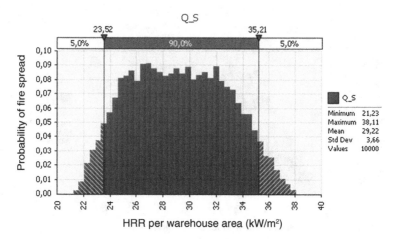

Fig. 5.5 Distribution of fire spread potential as a function of radiation level in the southerly direction from the burning warehouse. Graph was generated using @Risk software

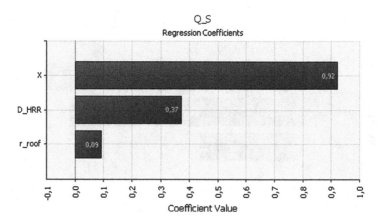

Fig. 5.6 Sensitivity of X, hazard occupancy class (D_HRR), and hole size (r_hole) to the fire spread potential in the southerly direction. Graph was generated using @Risk software

parameter. The negative sign indicates that the position of the hole is unfavorable to an increase in the output, Q_E. The opposite situation is true for Q_W.

The probability of fire spread results shown in Fig. 5.3 indicate that the upper limit of 33 kW/m^2 (3.1 kW/ft^2), which is relevant for the northerly direction, is never reached (the area under the curve above this limit is zero). There is 90 % confidence that the mean incident radiation is between 3.933 and 6.724 kW/m^2.

In the northerly and southerly directions, there is no sensitivity to the location of the hole in the roof because the hole in Case Study 1 was situated along the south side, therefore this input parameter is not included in Fig. 5.4. The magnitudes of

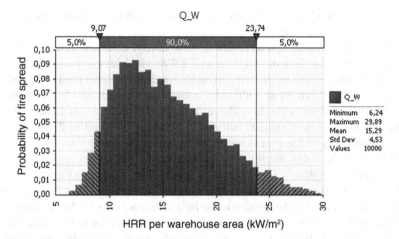

Fig. 5.7 Distribution of fire spread potential as a function of radiation level in the westerly direction from the burning warehouse. Graph was generated using @Risk software

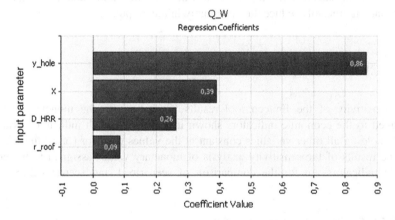

Fig. 5.8 Sensitivity of hole location (*y_hole*), X, hazard occupancy class (*D_HRR*), and hole size (*r_hole*) to the fire spread potential in the westerly direction. Graph was generated using @Risk software

the other input parameter sensitivities maintain the same order as the easterly and westerly directions but are much higher than in the easterly and westerly directions.

X, the ratio of radiated to total HRR, becomes the most sensitive parameter in the northerly and southerly directions. X is difficult to know accurately because warehouse fires are far from repeatable laboratory environments. Therefore, it would be difficult to reduce the uncertainty by improving the quality of the X estimation. The model is not very sensitive to the size of the hole (*r_roof*), which allows some flexibility in modelling the incident when the hole size is at its worst stage. The interval chosen for this case, 15–30 m (49–98 ft), represents a size that is within the worst case range.

The probability of fire spread results in the southerly direction shown in Fig. 5.5 indicate that the lower limit of 18 kW/m^2 (1.7 kW/ft^2) is always surpassed, i.e. there is 100 % probability of fire spread with favorable winds. The higher limit of 33 kW/m^2 (3.1 kW/ft^2), which is relevant without favorable wind, presents a 20 % probability of fire spread (the area under the curve above the limit in Fig. 5.5). As seen There is 90 % confidence that the mean incident radiation is between 23.52 and 35.21 kW/m^2.

The probability of fire spread results indicate that the lower limit of 18 kW/m^2 (1.7 kW/ft^2) falls within the 90 % probability of the fire spread region while the higher limit of 33 kW/m^2 (3.1 kW/ft^2) is never reached. As seen in Fig. 5.7, there is 90 % confidence that the mean incident radiation is between 9.07 and 23.74 kW/m^2.

As shown in Fig. 5.8, the location of the hole in the roof is by far the most sensitive input parameter for fire spread in the westerly direction, similar to fire spread in the easterly direction. In this case, the positive sign indicates that the position of the hole is favorable to an increase in the output, Q_W. However, the position of the hole in the roof is treated as a stochastic parameter in the model that is difficult to know with any accuracy. If an exact hole location were known, a modified model could take this as an input and significantly reduce the uncertainty in the output.

5.2   CBA

The sensitivity of the Enveco tool results to changes in the monetary values assigned to the economic indicators shown in Table 5.1. Each indicator is varied while holding all other variables constant at the values used for Case Study 1.

The results of the sensitivity analysis of monetary values assigned to the economic indicators are for the comparison of scenario 1 and scenario 2. Strong

Table 5.1 Sensitivity of economic indicators

Indicator	Range of input		Results
	Low	High	% change output/% change input
Firefighter fatalities	$4.5 million	$18 million	3.0
Firefighter injuries	$4000	$16000	0
Replace damaged property	$100/m^2	$400/m^2	0.3
Fire service response	$300/truck/h	$1000/truck/h	See overall sensitivity analysis (Table 5.4)
Job disruption	$19000/year	$39000/year	0.2
Direct business interruption	10 %	25 %	See overall sensitivity analysis (Table 5.4)
Indirect business interruption	5 %	20 %	0
Rent reduction	$2/m^2	$10/m^2	0

sensitivities to some of the input data don't necessarily appear when the comparison is made and therefore do not affect the results as much as one might expect. In fact, the firefighter fatality indicator is the only one that has a significant effect on the results. This is because firefighter fatalities are only considered in scenario 1 and thus do not cancel out of the comparison. The results are expressed as absolute value of the ratio of percent change in tool output to percent change in tool input, meaning a high number indicates a high sensitivity.

Property damage and job disruption also have a measurable effect on the sensitivity results because they are related to the warehouse area, which differs between scenario 1 and scenario 2. The results for these indicators are both <0, showing that the relationship on warehouse area has a dampening effect on the sensitivity.

5.3 LCA

5.3.1 Sensitivity of LCA Model Input

The sensitivity of the LCA models to changes in input parameters can be determined by running the models using different values for the parameters and observing the differences in the results. There are two models: the Athena Impact Estimator for Buildings, which is used for replacement of the warehouse structure due to fire damage; and SimaPro, which is used to estimate the impacts of replacing the damaged contents, the fire effluent, and the fire response.

It was assumed that the MBS system of construction was suitable for the model warehouse. The sensitivity of the model results to the materials used to construct the warehouse is shown in terms of percent change from the MBS results in Table 5.2.

Table 5.2 Sensitivity of warehouse construction materials

Impact category	MBS	SIP	Concrete	% change SIP	% change concrete
Global warming potential (kg CO_2 eq)	1957179	1817394	1991284	7.1	1.7
Acidification potential (kg SO_2 eq)	8383	7848	8601	6.4	2.6
Eutrophication potential (kg N eq)	597	584	6431	2.2	7.7
Ozone depletion potential (kg CFC-11 eq)	0.02	0.02	0.02	4.5	11.8
Smog potential (kg O_3 eq)	123656	115967	127936	6.2	3.5
Energy (MJ)	23285856	21649534	22914737	7.0	1.6

Note that the Athena Impact Estimator for Buildings does not include the Ecotoxicity impact category

For the sensitivity analysis two other warehouse models were created, one model uses structural insulated panels (SIP) and the other model uses tilt-up concrete panels for the walls. The roof, columns and beams, foundation, and openings are the same in all three warehouse models.

The results in Table 5.2 show that the impact categories are generally not very sensitive to the wall materials used to construct the warehouse. The largest percent change is less than 12 % and the average percent change is 5.2 %.

The sensitivity of the results of the LCA models used for the contents, fire effluents, and fire service response was explored using the SimaPro LCA software. For the contents, the mix of contents was varied randomly 5 times and the results compared with the default mix of contents. The results are shown in Table 5.3.

It can be seen in Table 5.3 that there is considerable sensitivity in the LCA model to the mix of contents stored in the warehouse. Several of the results changed more than 80 %. Considering that replacing the contents is the largest source of impact in four of the seven impact categories, this high level of sensitivity is a serious concern.

The fire effluents were selected based on previous work in which furnished rooms or non-industrial structures were burned [47, 48]. The fire response consists of driving the fire trucks to and from the incident and possibly includes contamination of soil and water from fire water run-off. Both of these sources contribute many orders of magnitude less impact than replacing the warehouse structure and contents, therefore it is not useful to perform a sensitivity analysis specifically on these models.

Table 5.3 Results of random variations in the mix of warehouse contents

Impact category	Original mix	% change test 1	% change test 2	% change test 3	% change test 4	% change test 5
Global warming potential (kg CO_2 eq)	22622	11.2	21.8	23.1	73.2	2.5
Acidification potential (kg SO_2 eq)	30996	7.1	21.2	29.1	78.0	1.4
Eutrophication potential (kg N eq)	920433	5.6	20.0	46.8	91.1	1.0
Ozone depletion potential (kg CFC-11 eq)	93556	86.8	33.6	77.5	7.1	18.9
Smog potential (kg O_3 eq)	26199	3.4	20.7	33.8	81.8	0.6
Ecotoxicity (CTUe eq)	6409033	8.3	20.1	50.0	93.7	1.3
Energy (MJ)	5274720	54.4	12.9	5.0	87.8	22.3

Table 5.4 Highlights of overall sensitivity analysis of Enveco tool comparison results to input parameters. Only results greater than or equal to 1 are shown

Input	Unit	Input value range		Results
		Low	High	
Distance to threatened structures	m	100	400	No results ≥1
Roof hole size	m	7–15	30–45	
Hole location	–	Far	Near	
Hazard group	–	Light	Extra-Group 2	Property damage, rent reduction, and all environmental results = 138. Job disruption and direct business interruption = 332. Total economic savings = 64.7
Compartment size % of warehouse size	%	20	80	Property damage, rent reduction, and all environmental results = 2.64. Job disruption and direct business interruption = 5.75. Total economic savings = 5.40
Warehouse structure area	%	48	192	Property damage, rent reduction, job disruption, direct business interruption and all environmental results = 1.01. Total economic savings = 1.06
Rack height	m	2	8	Eutrophication = 1.00. Ozone depletion and ecotoxicity = 1.01
Burned area—structure	%	25	100	No results ≥1
Damaged area—structure	%	25	100	Property damage, business interruption, and rent reduction = 1.0
Unemployment	m²/person	30	80	No results ≥1
Business interruption % of property damage	%	10	25	
Fill factor	%	25	75	
Burned area—contents	%	25	100	
Damaged area—contents	%	25	100	Eutrophication, ozone depletion and ecotoxicity = 1.0
Mix of contents				See Table 5.3

(continued)

Table 5.4 (continued)

Input	Unit	Input value range		Results
		Low	High	
Cost of response	USD/truck/h	300	1000	No results ≥ 1
Number of trucks	Each	6	20	
Distance fire station to incident	km	5	30	
Amount of water into environment	L	1×10^6	1×10^7	
Number of firefighter fatalities	Each	0	5	Firefighter fatalities = 4.5×10^7, total economic savings = 5.0
Number of firefighter injuries	Each	0	20	Firefighter injuries = 1.7×10^5

5.3.2 Uncertainty of LCA Model Results

The Athena Impact Estimator for Buildings claims an uncertainty of around 15 %, which can be applied to the environmental impact of replacing the damaged warehouse structure [23]. The uncertainty of the LCA results for replacing the burned or damaged contents and the impact of the fire effluent and fire service response is determined using a built-in Monte Carlo analysis with a confidence interval of 95 % and 1000 iterations (different criteria than those used in the QRA uncertainty analysis). The results are shown in Fig. 5.9 for 1 kg of each category of warehouse contents (aggregated), fire emissions to air from 1 m^2 of ware-house + contents burned, fire emissions to water from 1 m^3 of fire water run-off, and 1 fire truck carrying 3.8 m^3 (1000 gal) of water. The impacts of replacing the warehouse are also shown.

The error bars shown in Fig. 5.9 were generated by the Monte Carlo analysis and reflect the ability of the LCA model to produce consistent results, with the exception of the error bars on the warehouse results, which are ± 15 % according to the Athena user manual [23]. The distribution of possible data values is assumed to be lognormal. The uncertainty related to knowledge of the data quality is included in the analysis via the variance (the spread of the lognormal distribution). The variance is estimated using a tool called "Pedigree" that is based on these factors: data reliability, completeness, temporal correlation, geographical correlation, and further technological correlation.

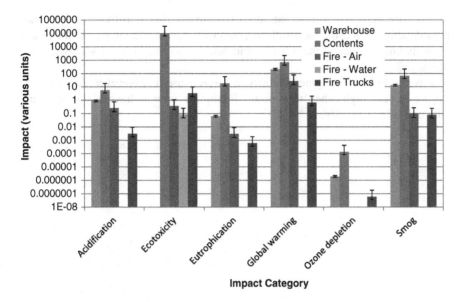

Fig. 5.9 Uncertainty of LCA model of warehouse contents, fire emissions to air and water, and fire trucks used in response. Note that the y-axis is logarithmic

5.4   Overall Enveco Tool Sensitivity

Many of the input data are used in more than one component of the Enveco tool, making it difficult to assess the sensitivity of the tool results by examining the sensitivity of the three components of the tool separately. An overall sensitivity analysis was performed on the input data with respect to their effect on the savings due to fire service intervention, i.e. only the input that contributes to a difference between scenario 1 and scenario 2. The models used in the tool have strong sensitivities to some of the input data that don't appear when the comparison is made and therefore do not affect the results as much as one might expect. The results are expressed as absolute value of the ratio of percent change in tool output to percent change in tool input, meaning a high number indicates a high sensitivity.

Based on these results, the hazard group has the largest impact on the economic results. This is consistent with the QRA sensitivity analysis, where X, the ratio between radiated and total HRR, was found to be the most sensitive input parameter.

The next most impactful input parameters are the fully developed fire compartment size as a percent of the warehouse size (and the stand-alone warehouse size), both of which also produce significant environmental results. The storage rack height and the area of damaged contents represent the amount of contents that need to be replaced due to fire damage. As seen in Table 5.3, the environmental impacts are very sensitive to the mix of contents. This sensitivity is dampened considerably when using the results in a comparison, where the impact of the mix of contents only applies to the difference in content volume between scenario 1 and scenario 2.

It is rare to have firefighter fatalities in a warehouse fire, however, the economic impact is very high when they occur. Firefighter injuries have a similar result, although to a lesser degree.

Chapter 6
Case Studies

Use of the Enveco tool is demonstrated on two case studies in the following sections. Case Study 1 was selected because it was surrounded by industrial structures on all sides, the fire vented through the roof, and the fire service adopted a defensive strategy using water to cool the adjacent structures. The hazard occupancy was classified as "Extra—Group 1" due to the mix of contents, some of which could affect the burning behavior of the fire. Therefore another case study was selected as a second opportunity to demonstrate the Enveco tool. This warehouse had only one adjacent structure (also industrial) and the contents were classified as "Ordinary—Group 1".

See Appendix B for additional background information about analysis of historic warehouse fires as potential case studies.

6.1 Case Study 1

A warehouse fire completely destroyed a 7686 m² (122 m × 63 m) (83,730 sf, 400 ft × 208 ft) multi-tenanted warehouse. The fire began in mid-afternoon in the southern portion of the warehouse and rapidly spread within the Tenant 1 part of the building having an area of 4221 m² (67 m × 63 m) (45,434 sf, 220 ft × 208 ft). It also spread to stored material outside the southeast corner of the building, see Figs. 6.1 and 6.2. The fire service arrived when the fire started to become fully developed and decided to battle the fire defensively. The fire self-vented through the roof with debris being thrown up in the air in a thermal column. The fire service extinguished exposure fires by spraying water on surrounding surfaces. Next, the south wall and parts of the west wall collapsed. An air conditioning facility about 23 m (75 ft) south of the building was under threat from the fire and was being cooled by the fire service.

After almost two hours from the fire outbreak the solid concrete wall that separated the burning part occupied by Tenant 1 from the other part occupied by

© Fire Protection Research Foundation 2016
F. Amon et al., *Development of an Environmental and Economic
Assessment Tool (Enveco Tool) for Fire Events*, SpringerBriefs in Fire,
DOI 10.1007/978-1-4939-6559-5_6

Fig. 6.1 Google map of the incident location today (from Dec 2015). Based on this information the distance to the east (plastic) structure is assumed to be the same as the distance to the west structure (32 m or 105 ft)

Tenant 2 collapsed and the fire continued to spread through the north section of the warehouse. With a large portion of the north roof being intact it was difficult to reach the seat of the fire. After several hours a larger part of the roof collapsed and it was easier to reach the seat of the fire of the north section and the fire could be controlled. By the next morning the entire building and contents were destroyed.

The ceiling height was approximately 9 m (30 ft). The building was equipped with an automatic sprinkler system throughout, which was not effective for the mix of contents in the warehouse. The average temperature on the day of the incident was 33 °C (91 °F) with winds ranging from 5 to 13 km/h (3–8 mph), direction unknown. Considering that there was a risk of fire spread to the air conditioning facility, the wind was supposedly blowing from north to south, although the east to west direction is more common for this location in August, as shown in Fig. 6.3. The photo of the fire scene in the fire investigation report [51] indicates that the wind direction could also have been from west to east, which is also supported by fire service cooling of the plastics manufacturing facility shown in Fig. 3.2 of the fire investigation report and in Figs. 6.1 and 6.2.

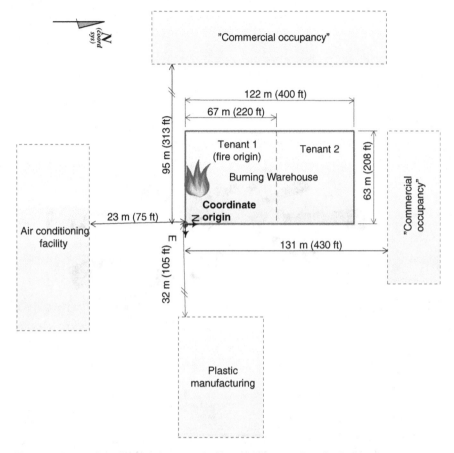

Fig. 6.2 Sketch of burning warehouse and adjacent structures, Case Study 1

6.1.1 Quantitative Risk Assessment

This section shows how the proposed QRA model outlined in Sect. 3.1 can be applied to Case Study 1 to estimate the risk of fire spread without fire service intervention. First a deterministic case is given to show how the model works. The model is applied in a spreadsheet, although the sensitivity analysis in Sect. 5.1 was conducted using @Risk software. The equations in this section were originally designed to be used with SI units, therefore the calculations and associated discussion are presented in SI units.

6.1.1.1 Fire Spread—Deterministic Example Calculations

For estimating the flame height and radiation to adjacent buildings the maximum HRR must be estimated. Based on the fire investigation report the worst phase was

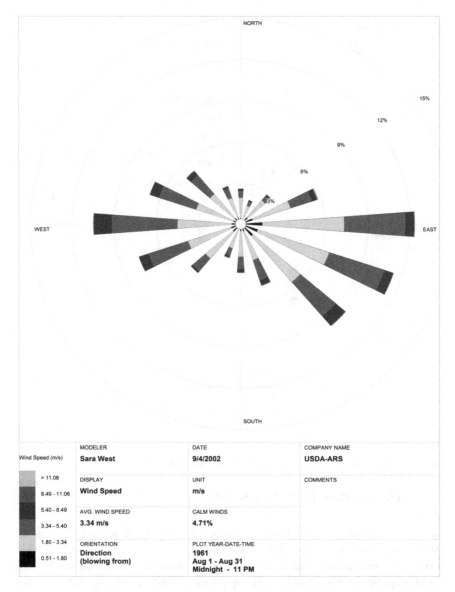

Fig. 6.3 Wind directions for Case Study 1 in August. http://www.wcc.nrcs.usda.gov/ftpref/
downloads/climate/windrose/arizona/phoenix/

when the fire became fully developed in the Tenant 1 part of the warehouse. Based
on the content that was burning and estimates on previous fully developed warehouse fires and their flame lengths [28], a HRR ranging from 0.5 to 1.3 MW/m^2 can
be expected. The lower number is for solid fuels such as wood furniture and the
higher number is for hydrocarbon pool fires. Since this fire included both a mixture

of fuels, 1 MW/m^2 was chosen. From the fire investigation report [52] it can be inferred that the whole Tenant 1 part of the warehouse was involved in a fully developed fire and the roof caved in before the fire rated wall to the Tenant 2 part of the warehouse collapsed. These assumptions result in a total HRR of 4 GW. Assuming the diameter of the hole in the roof to be 20 m, the flame length becomes:

$$H_f = -1.02 * 20 + 0.235 * 4,000,000^{\frac{2}{5}} = 82 \text{ m}$$

For comparison, a diameter of 5 m results in 97 m flames and a diameter of 50 m results in 52 m flames.

Assuming a wind of 4 m/s and an average $H_c = 30$ MJ/kg, $\dot{m}''$, the mass burning rate through the hole is estimated as:

$$\dot{m}'' = \frac{\dot{Q}}{XAH_c} = \frac{4,000,000}{0.3\pi 10^2 30,000} = 1.4 \text{ kgs}^{-1} \text{ m}^2$$

which results in u^*:

$$u^* = \frac{u}{\left(\frac{g\dot{m}''D}{\rho}\right)^{\frac{1}{3}}} = \frac{4}{\left(\frac{9.82*1.4*20}{1,2}\right)^{\frac{1}{3}}} = 0.65$$

u^* is less than 1, meaning that there is no or limited flame tilt. Assuming an adjacent structure with the same roof height as the burning warehouse and the hole in the roof is situated towards the south side of the roof, the distance to the flame center becomes $\sqrt{33^2 + 41^2} = 52$ m. Assuming $X = 0.3$, the radiation becomes:

$$\dot{Q}'' = X\frac{\dot{Q}}{4\pi L^2} = 0.3\frac{4,000,000}{4\pi 52^2} = 35 \text{ kW/m}^2$$

Considering flying embers in the air and that many materials, e.g. wood, ignite at 10–14 kW/m^2 for piloted ignition, it seems likely that the fire could have spread to the air-conditioning facility, and to the plastic manufacturing facility that were located at a similar distance. If the hole in the roof is situated towards the west side, the radiation to the commercial occupancy in the west becomes ($L = \sqrt{42^2 + 41^2} = 59$ m):

$$\dot{Q}'' = X\frac{\dot{Q}}{4\pi L^2} = 0.3\frac{4,000,000}{4\pi 59^2} = 27 \text{ kW/m}^2$$

This means that fire spread could have occurred to the building to the west of the burning warehouse as well. The radiation towards the commercial occupancy in the north, assuming the roof is intact on the whole Tenant 2 part of the warehouse, and a hole exists in the roof towards the north end of the garden supply warehouse, results in ($L = \sqrt{70^2 + 41^2} = 81$ m):

$$\dot{Q}'' = X\frac{\dot{Q}}{4\pi L^2} = 0.3\frac{4,000,000}{4\pi 81^2} = 15 \text{ kW/m}^2$$

At this distance ignition can only take place through piloted ignition, which can happen if burning embers fly 60 m towards the north. If the wind direction were from north to south or from west to east, ignition of the north occupancy is not likely to occur from the garden supply fire. As the fire spread to the Tenant 2 part of the warehouse the risk increased, however, the hole in the roof towards the south was unfavorable for fire spread in the northerly direction.

6.1.1.2 Risk Simulations

There are both known and uncertain input parameters in the previous analysis. For example it is known from the investigation report that the whole warehouse burned to the ground resulting in a total loss. This would have been the case with or without fire service intervention. It is also known from the investigation report that the fire did not spread to any other buildings and that the fire service cooled some of the adjacent structure surfaces. It is uncertain whether the fire would have spread to adjacent structures without fire service intervention. Key uncertainties and assumptions in the probabilistic model are described in Sect. 3.1.2.

The QRA input parameters are listed below. Parameters can be input to the Enveco tool either in British units or SI units; British units will be used in the following example.

- Distances from coordinate system origin to threatened structures:

  - North: 430 ft
  - East: 313 ft
  - South: 75 ft
  - West: 105 ft

- The wind direction is assumed to be roughly 70 % from west and 30 % from north based on the wind rose shown in Fig. 6.3 and the incident investigation report. It is therefore assumed that the east and south buildings had a 70 and 30 % probability of being threatened by flying embers, respectively. The wind direction will affect the likelihood for ignition (cold or warm wind with flying embers).
- The initial location and diameter of the hole in the roof is inferred to be in the south side of the roof, over the location where the fire started. In most cases the roof will first be weakened above where the fire starts. The worst case diameter for the hole is between 49 and 98 ft ($< \frac{1}{2}L$), evenly distributed along the south side of the roof.

Warehouse	Probability, saved property (%)
South	46
North	0
West	0
East	25
Sum	72

Table 6.1 Estimated percentage of saved property due to prevention of fire spread to adjacent structures

- Considering the height, amount, and mixture of commodities stored, hazard occupancy class 'Extra—Group 1' seems most suitable to describe the fire load per square feet.
- The fire wall lasted for about 2 h between the Tenant 1 and Tenant 2 parts of the warehouse. The roof caved in before the Tenant 2 part was involved in the fire. Therefore the structure width, length and area used for the risk of fire spread is based on the fully developed fire in the Tenant 1 part of the warehouse. This fire compartment is 220 ft × 208 ft.

Monte Carlo simulations are run in a spreadsheet with the input and model described above to calculate the risk of fire spread to adjacent structures, see resulting output in Table 6.1.

The probabilities are summed to account for the wind vector, which appeared to be blowing from the west-northwest. Hence, when combined with the proximity of the adjacent structures, results in risks in both the southerly and westerly structures. Under the assumption that the four adjacent warehouses were of the same occupancy and size as the burning warehouse, the fire department intervention results in a property savings of 72 %.

6.1.2 Cost Benefit Analysis

The practical steps for a CBA presented in Sect. 3.2 are applied in this section to Case Study 1.

1. **Define scope and comparison scenarios**: this is an analysis of a fire that occurred in the warehouse described above. This analysis compares the true scenario that the fire service responded to the fire with the theoretical scenario in which the fire service did not respond.
2. **Specify who will use the results**: the users were originally intended to be the fire service, however, other stakeholders such as community administrators, warehouse owners or managers, insurance companies, etc. may also find value in using it.
3. **Identify the indicators to be measured**: the indicators are listed under point 5, along with the input values for the CBA calculations.
4. **Assess the economic consequences**: the results of the CBA are presented in Table 6.2, along with the results of the QRA and LCA.

Table 6.2 Input and output of Enveco tool for Case Study 1

Value	Units	Default	Parameter
Input			
Value	Units	Default	Parameter
Risk of fire spread			
Distance from coordinate origin to threatened structure(s)			
430	ft	n/a	North
313	ft	n/a	East
75	ft	n/a	South
105	ft	n/a	West
Direction wind is coming from			
30	%	25	North
0	%	25	East
0	%	25	South
70	%	25	West
Range of possible sizes of hole in roof			
49	ft	49	Smaller diameter
98	ft	98	Larger diameter
Approximate location of hole in roof			
South		Anywhere	Hole location
Hazard occupancy class			
Extra—Group 1		Ordinary—Group 2	From NFPA 13
Warehouse Description			
Fully developed fire compartment			
220	ft	0	x-dimension
208	ft	0	y-dimension
Structure area			
400	ft	n/a	x-dimension
208	ft	n/a	y-dimension
Storage rack height			
20	ft	20	Rack height
Sprinklers			
0.3	gpm	0.3	Sprinkler water
20	Min	n/a	Sprinkler activation time
0	%	0	Sprinkler water run-off
Fire damage with fire service intervention (known damage)			
100	%	100	Burned structure area
100	%	100	Damaged structure area
268	Each	268	People unemployed
379,827	USD	379,827	Business interruption
Contents			
50	%	50	Fill factor
100	%	100	Burned contents area
100	%	100	Damaged contents area

(continued)

Table 6.2 (continued)

Input			
Value	Units	Default	Parameter
10	%	10	Plastic
15	%	15	Paper
15	%	15	Wood
10	%	10	Textile
15	%	15	Soft furniture
10	%	10	Chemicals
15	%	15	Dry food
10	%	10	Electronics
Fire service response			
600	USD	600	Cost of response
6	#	6	Trucks
6	Miles	6	Distance travelled
12	h	12	Time at incident
1,620,000	Gal	1,620,000	Water used
10	%	10	Water to environment
0	Person	0	Firefighter fatalities
0	Person	0	Firefighter injuries
Output (comparison of scenario 1—scenario 2)			

Savings due to fire service response—QRA

On average:	**0.72**	**Warehouses**

Savings—economic assessment

Value	*Unit*	*Description*
0	USD	Firefighter fatalities
0	USD	Firefighter injuries
927,366	USD	Property damage
1,804,068	USD	Job disruption
271,018	USD	Direct business interruption
27,102	USD	Indirect business interruption
−43,200	USD	Fire service intervention
23,932	USD	Rent reduction
3,010,287	USD	Total economic savings

Savings—environmental assessment

	Fire emissions	Structure replaced	Contents replaced	Fire department	Total Savings
Global warming (kg CO_2 equiv)	6	1,170,982	16,267	−3.3E−03	1.2E+06
Acidification (kg SO_2 equiv)	11	5015	22,289	−2.7E−03	2.7E + 04
Eutrophication (kg N equiv)	1.3	357	661,873	−5.4E−03	6.6E+05

(continued)

Table 6.2   (continued)

Output (comparison of scenario 1—scenario 2)					
Ozone depletion (kg CFC-11 equiv)	0	0	67,275	−1.7E−03	6.7E+04
Smog (kg O_3 equiv)	0.3	73,984	18,840	−5.1E−03	9.3E+04
Ecotoxicity (CTUe equiv)	0.2	n/a	4,608,659	−3.6E−02	4.6E+06
Energy used (MJ)	n/a	13,931,948	3,792,988	n/a	1.8E+07

5. **Use discount rate and cash flow analysis to calculate net present value (NPV)**: the US Consumer Price Index (CPI) was used to calculate NPV whenever source data was not current. All economic values are in 2016 United States dollars (USD).

   (a) Firefighter fatalities ($9,000,000/life)
   (b) Firefighter injuries ($8257/injury)
   (c) Replace damaged property ($166.8/m^2)
   (d) Fire service response ($600/truck/h)
   (e) Job disruption ($16,229/person)
   (f) Direct business interruption ($380,129)
   (g) Indirect business interruption ($38,013)
   (h) Rent reduction ($4.31/m^2)

6. **Sensitivity analysis of the results**: see Chap. 5, Sensitivity and Uncertainty Analyses.
7. **Recommend specific action**: see Chap. 7, Conclusions, and Chap. 8, Future Work.

6.1.3   Life Cycle Assessment

This application of the LCA to the case draws mostly on input of the warehouse structural and content specifications, for example, input related to the size of the warehouse and amount of burned and damaged material. These default values were used for determining the types of materials present as contents requiring replacement due to the fire (in % of total contents):

- Plastic (10 %)
- Paper (15 %)
- Wood (15 %)
- Textile (10 %)
- Soft furniture (15 %)

- Chemicals (10 %)
- Dry food (15 %)
- Electronics (10 %).

For the fire service response, 6 trucks responded and used an estimated 1.62 million gallons of water, 10 % of which was assumed to escape into the environment.

6.2   Case Study 2

The first emergency (911) call, received shortly before 7:00 am, reported a huge fire at an industrial park. According to the NFPA fire investigator [53], the warehouse structure measured 600 ft north to south and 1000 ft east to west and was approximately 30 ft tall. The major part of the building was used as a warehouse, with about one sixth of the property occupied as offices at the eastern end. In the north there was an annex of approximately 100 ft by 520 ft. The fire started in the southwest corner of the warehouse and was eventually stopped at the eastern portion of the structure near the office area and at the warehouse annex at the north side. In the north there was an adjacent warehouse structure of similar size to the burned structure, see Fig. 6.4. The warehouse mainly stored parts for appliances with rack heights between 12 and 24 ft, having a classification of IV in NFPA 13 [53, 54].

6.2.1   Quantitative Risk Assessment

This section shows how the proposed QRA model outlined in Sect. 3.1 can be applied to Case Study 2 to estimate the risk of fire spread without fire service intervention. The fire department intervention was successful in stopping the fire from spreading from the fully developed fire compartment to the east and north. Without fire department intervention, it is assumed that such a large fire would have included the whole structure leading to a total loss of Building 6, including the office space in the east and the annex in the north. The QRA model is used to estimate the added risk from fire spread to the warehouse structure in the north (Building 5). Distances and an overview of the structures can be seen in Fig. 6.5.

The wind conditions during the fire were such that the fire plume, according to pictures of the fire scene [53] was blown above the adjacent warehouse structure (Building 5), from the south to the east. The wind speed during the main risk of fire spread (from 7:00 to 11:00) was between 6.7 and 8.9 m/s (15–20 mph) according to weather reports on the day of the incident. This would result in an added risk due to the tilting of the flame, but this added risk is excluded since it is not yet implemented in the tool. Since the fire started in the southwest corner of the warehouse

Fig. 6.4 Google map
of the incident location,
Case Study 2

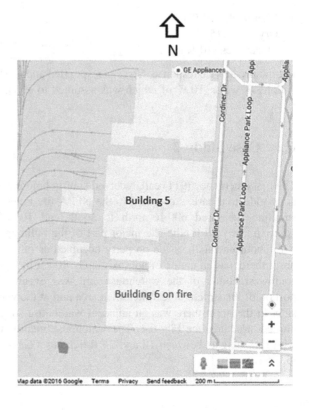

Fig. 6.5 Input data used for
the QRA model to estimate
the risk of fire spread to the
north warehouse

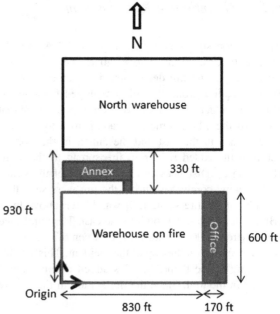

and was stopped at the east office space and at the annex in the north, the roof collapse that is mentioned in radio transcripts at 07:19 can be expected to have been on the south side, furthest away from the north warehouse structure. The later roof collapse at 12:20 can be expected to have led to a mass fire with a limited potential for larger flames.

6.2.2   Risk Simulations

For the Case Study 2 fire the input parameters are listed below. Parameters can be input to the Enveco tool either in British units or SI units; British units will be used in the following example.

- The warehouse building dimensions were:
  - Length: 830 ft
  - Width: 600 ft
- The minimum distance between the warehouse on fire and adjacent north structure was 103 ft (from Google Earth images). Distances from the coordinate system origin to threatened structures become:
  - North: 930 ft
  - East: (n/a)
  - South: (n/a)
  - West: (n/a)
- The wind direction was from the south, such that the north structure was threatened by flying embers according to pictures of the fire [53].
- The location of the hole in the roof highly affects the risk of fire spread. In this case the fire started in the southwest corner [54] and, based on photos [53, 54], it seems the hole in the roof was along the south side of the warehouse. The worst case diameter for the hole is uncertain. A large interval is therefore selected between 50 and 400 ft ($< \frac{1}{2}L$).
- Considering the type and in particular the height of commodities stored (up to 24 ft), hazard occupancy class 'Extra—Group 1' seems most suitable to describe the fire load per square feet [53].
- The fire is assumed to have involved the entire warehouse part of the structure at the time of the largest risk for fire spread (between 07:19 and 12:20).

Monte Carlo simulations are run in a spreadsheet with the input and model described above to calculate the risk of fire spread to the adjacent structure in the north. Under the assumption that the adjacent warehouse was of the same occupancy as the burning warehouse, the fire department intervention results in property savings of 8.3 % despite the fact that the hole in the roof was situated on the south side furthest away. The calculation is very sensitive to the distance between the buildings and the positioning and size of the hole in the roof. The distance between the buildings was estimated from Google maps and although uncertain, the true value should be close to the estimated value. Based on the incident report it seems quite sure that the hole was on the south side, but the size of the hole is more uncertain. A lower hole diameter of 300 ft results in a fire spread in 25 % of the cases and a hole diameter between 50 and 100 ft results in 0 % fire spread. The initial estimate of 8.3 % should be used unless more information could be found to give a more precise estimate of the hole diameter.

6.2.3   Cost Benefit Analysis

The practical steps for a CBA presented in Sect. 3.2 are applied in this section to Case Study 2. The monetary values used for this case study are the same as those used for Case Study 1.

1. **Define scope and comparison scenarios**: this is an analysis of a fire that occurred in the warehouse described above. This analysis compares the true scenario that the fire service responded to the fire with the theoretical scenario in which the fire service did not respond.
2. **Specify who will use the results**: the users were originally intended to be the fire service, however, other stakeholders such as community administrators, warehouse owners or managers, insurance companies, etc. may also find value in using it.
3. **Identify the indicators to be measured**: the indicators are listed under point 5, along with the input values for the CBA calculations.
4. **Assess the economic consequences**: the results of the CBA are presented in Table 6.3, along with the results of the QRA and LCA.
5. **Use discount rate and cash flow analysis to calculate net present value (NPV)**: the US Consumer Price Index (CPI) was used to calculate NPV whenever source data was not current. All economic values are in 2016 United States dollars (USD).

   (a) Firefighter fatalities ($9,000,000/life)
   (b) Firefighter injuries ($8257/injury)
   (c) Replace damaged property ($166.8/m^2)
   (d) Fire service response ($600/truck/h)
   (e) Job disruption ($16,229/person)
   (f) Direct business interruption ($380,129)
   (g) Indirect business interruption ($38,013)
   (h) Rent reduction ($4.31/m^2)

6. **Sensitivity analysis of the results**: see Chap. 5, Sensitivity and Uncertainty Analyses
7. **Recommend specific action**: see Chap. 7, Conclusions, and Chap. 8, Future Work.

Table 6.3 Input and output of Enveco tool for Case Study 2

Value	Units	Default	Parameter
Input			
Risk of fire spread			
Distance from coordinate origin to threatened structure(s)			
930	ft	n/a	North
10,000	ft	n/a	East
10,000	ft	n/a	South
10,000	ft	n/a	West
Direction wind is coming from			
0	%	25	North
0	%	25	East
100	%	25	South
0	%	25	West
Range of possible sizes of hole in roof			
50	ft	49	Smaller diameter
400	ft	98	Larger diameter
Approximate location of hole in roof			
South		Anywhere	Hole location
Hazard occupancy class			
Extra—Group 1		Ordinary—Group 2	From NFPA 13
Warehouse description			
Fully developed fire compartment			
830	ft	0	x-dimension
600	ft	0	y-dimension
Structure area			
830	ft	n/a	x-dimension
600	ft	n/a	y-dimension
Storage rack height			
20	ft	20	Rack height
Sprinklers			
0.3	gpm	0.3	Sprinkler water
20	Min	n/a	Sprinkler activation time
0	%	0	Sprinkler water run-off
Fire damage with fire service intervention (known damage)			
100	%	100	Burned structure area
100	%	100	Damaged structure area
1001	each	1001	People unemployed
1,420,000	USD	1,420,000	Business interruption
Contents			
50	%	50	Fill factor
100	%	100	Burned contents area

(continued)

Table 6.3 (continued)

Input			
Value	Units	Default	Parameter
100	%	100	Damaged contents area
40	%	10	Plastic
30	%	15	Paper
10	%	15	Wood
0	%	10	Textile
0	%	15	Soft furniture
10	%	10	Chemicals
0	%	15	Dry food
10	%	10	Electronics
Fire service response			
600	USD	600	Cost of response
16	#	6	Trucks
6.2	Miles	6	Distance travelled
120	h	12	Time at incident
43,200,000	Gal	43,200,000	Water used
10	%	10	Water to environment
0	Person	0	Firefighter fatalities
0	Person	0	Firefighter injuries
Output (comparison of scenario 1—scenario 2)			
Savings due to fire service response—QRA			
On average:	**0.08**	**Warehouses**	
Savings—economic assessment			
Value	*Unit*	*Description*	
0	USD	Firefighter fatalities	
0	USD	Firefighter injuries	
633,687	USD	Property damage	
1,232,755	USD	Job disruption	
116,570	USD	Direct business interruption	
11,657	USD	Indirect business interruption	
−1,152,000	USD	Fire service intervention	
16,353	USD	Rent reduction	
859,022	USD	Total economic savings	

Savings—environmental assessment

	Fire emissions	Structure replaced	Contents replaced	Fire department	Total Savings
Global warming (kg CO_2 equiv)	4	800,155	9592	−9.0E−03	8.1E+05
Acidification (kg SO_2 equiv)	7	3427	13,487	−7.5E−03	1.7E+04

(continued)

Table 6.3   (continued)

Output (comparison of scenario 1—scenario 2)					
Eutrophication (kg N equiv)	0.9	244	440,078	−1.5E−02	4.4E+05
Ozone depletion (kg CFC-11 equiv)	0	0	2878	−4.7E−03	2.9E+03
Smog (kg O$_3$ equiv)	0.2	50,554	11,924	−1.4E−02	6.2E+04
Ecotoxicity (CTUe equiv)	0.1	n/a	3,127,502	−9.9E−02	3.1E+06
Energy used (MJ)	n/a	9,519,968	352,807	n/a	9.9E+06

6.2.4   Life Cycle Assessment

This application of the LCA to the case draws mostly on input of the warehouse structural and content specifications, for example, input related to the size of the warehouse and amount of burned and damaged material. These default values were used for determining the types of materials present as contents requiring replacement due to the fire (in % of total contents):

- Plastic (40 %)
- Paper (30 %)
- Wood (10 %)
- Textile (0 %)
- Soft furniture (0 %)
- Chemicals (10 %)
- Dry food (0 %)
- Electronics (10 %).

For the fire service response, 16 trucks responded and used an estimated 43.2 million gallons of water, 10 % of which escaped into the environment.

Chapter 7
Conclusions

The goal of this project was to determine whether it is feasible to create a tool that can provide useful, consistent information to the fire service about the environmental and economic savings associated with their operations. Specifically, this tool is intended to help the fire service understand and share information about environmental and economic savings to the communities they protect.

As a start, this feasibility study is applied only to fully developed warehouse fires located in industrial areas for which water was used as the fire suppression agent. These simplifications were imposed in this first phase of the tool to avoid complications that might obscure the feasibility of developing the tool. Plans for future expansion of the tool functionality are listed in Chap. 8, Future Work.

Even though there are many uncertainties associated with the components of the Enveco tool due to non-specific input data, many of them become less severe when used in the comparison between the theoretical baseline case of no fire service intervention and the real case where the fire service intervenes. This is because some portion of these uncertain parameters used in the models are applied to both scenarios in the comparison and are therefore partially cancelled out of the results.

The most important factor is X, the ratio between radiated and total HRR, which determines a large portion of the difference between the two scenarios in terms of fire spread to adjacent structures. Other important factors are those that are related to the differences between the two scenarios, such as the area used to calculate property damage. There are three factors that are only present in one of the scenarios: firefighter fatalities, firefighter injuries, and the cost of the fire service intervention. Of these, firefighter fatalities have a high impact on the results, but have little uncertainty.

The Enveco tool can be updated with more accurate information as it becomes available. Likewise, as more information becomes available to the users through public or private databases, the tool will become easier to use and the quality of the input data can improve. The functionality of the tool can also be extended more easily as new information becomes available.

© Fire Protection Research Foundation 2016
F. Amon et al., *Development of an Environmental and Economic Assessment Tool (Enveco Tool) for Fire Events*, SpringerBriefs in Fire, DOI 10.1007/978-1-4939-6559-5_7

Finally, it is possible to generate useful environmental and economic impact information through a relatively simple tool such as the Enveco tool. This tool was originally intended for use by the fire service but it may have value for other interested parties as well. It is hoped that users will share their thoughts and expertise regarding future improvements.

Chapter 8
Future Work

The following items have either been suggested by others during the development of this tool or are the result of discussions within the project team. These ideas are not in prioritized order. See Appendix C for statistical decision support for possible future expansion of the Enveco tool.

- Additional structure types (single family residences, commercial buildings, apartment buildings, mixed types, etc.) and locations (wildland area, suburban neighborhood, central business district, etc.) will be identified in the future. The inventory will include information related to the fire, for example the structure content and construction type.
- Allow comparisons of outcomes (predictions) when different levels of response are taken. Vary number of personnel and amount/types of equipment and use response time as a factor.
- When and if the methodology is expanded to include more structure types, sites, and situations it can be implemented on a larger scale, for example in the form of an online service accessible to the fire service. It can also be introduced to the major fire service organizations via NFPA channels such as the NFPA Annual Meeting.
- The feature to input several fires could be implemented in the future, however this would require special conditions for the QRA.
- The current model does not cover materials with extreme flame temperatures such as magnesium. It covers fires for which the radiated HRR lies in the range 0.2 and 0.3. In a future model the X factor, ratio of radiated heat, could be made into an input parameter.
- The wind speed can be added as an input parameter to include the effect of tilted flames. This will be important for strong winds.
- A more advanced CBA with the fire station lifetime and NPV calculation based on several interventions could be implemented.
- Including costs of health degradation due to smoke and particles based on established studies could also be implemented.

© Fire Protection Research Foundation 2016
F. Amon et al., *Development of an Environmental and Economic
Assessment Tool (Enveco Tool) for Fire Events*, SpringerBriefs in Fire,
DOI 10.1007/978-1-4939-6559-5_8

- Include firefighter jobs and jobs created by the competitors of the business during the rebuilding phase.
- Possibly expand assumptions to address intentionally set versus accidental fires. Capture the value of fire investigative efforts.
- Include the impact/value of fire department salvage and overhaul operations.
- This study focused on a 'Defensive' fire response, future analysis would include offensive fires and successful interventions that will result in financial and environmental impacts within the primary fire occupancy as well.
- Distinguish between career and volunteer responses or assume something about the mix.

Appendix A
Existing Risk, Environmental, and Economic Assessment Models

by Mai Tomida and Brian Meacham
Worcester Polytechnic Institute
Department of Fire Protection Engineering
50 Prescott Street
Worcester, MA 01609

A.1 Life Cycle Assessment

Life Cycle Assessment (LCA) is a versatile tool to investigate the environmental impact of a product, a process or an activity by identifying and quantifying energy and material flows for the system [1]. The use of a product or a process involves more than just the production of the product or use of the process. Every single industrial activity is a complex network of activities that involve many different parts of society. Therefore, the need for a system perspective rather than a single object perspective has become vital in environmental research. The entire system has to be considered.

There are different computer software solutions for LCA calculations. Generally, the software can be divided into two different groups:

Specific LCA programs, (KCL-ECO, LCA Inventory Tool, SimaPro, etc.).

General calculation programs such as different spread sheet programs (Excel, etc.).

An LCA usually evaluates the environmental situation based on ecological effects and resource use. In a few cases the work environment has also been included. A traditional LCA does not cover the economic or social effects. A LCA only describes a process during normal operation. Accidents and other abnormal conditions are left out of the analysis usually due to lack of a consistent methodology or relevant data.

© Fire Protection Research Foundation 2016   67
F. Amon et al., *Development of an Environmental and Economic Assessment Tool (Enveco Tool) for Fire Events*, SpringerBriefs in Fire, DOI 10.1007/978-1-4939-6559-5

A.1.1 Fire-LCA

The Fire-LCA provides a good starting point for a holistic interpretation of a realistic life-cycle of a product including information concerning the probability that the product may be involved in a fire [1]. Though it is useful, it does not provide information concerning, for example, the effect of the toxicity of chemicals used in the product, number of lives saved, costs associated with the different cases

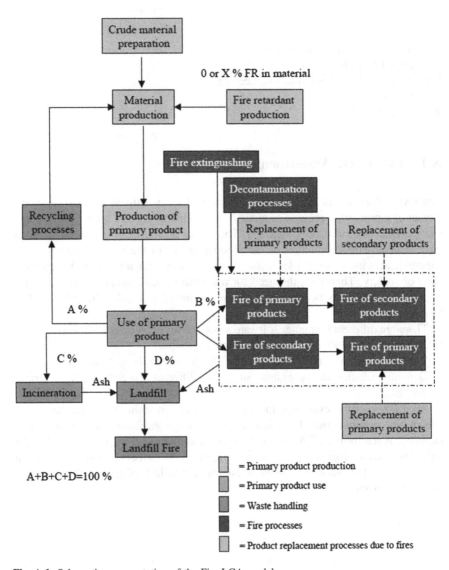

Fig. A.1 Schematic representation of the Fire-LCA model

or the societal effect of manufacturing practice. The emissions from fires contribute greatly to the environmental impact.

The Fire-LCA process is primarily the same process as the typical LCA process with the difference being that it includes modules to account for fire accidents. This methodology was developed at SP in Sweden. Figure A.1 shows a schematic representations of the Fire-LCA model. The aim of the model is to obtain a measure of the environmental impact of the choice of a given level of fire safety.

In a Fire-LCA model where the fire performance of a product or a process is evaluated, the actual function of the fire protection system could be how well the fire protection works or the number of fire occurrences for a given fire protection system [1].

A.2   Risk Assessment (RA)

The purpose of fire risk assessment (FRA) is to identify and characterize the fire risks of concern and provide information for fire risk management decisions [4]. The assessment is used to identify what can possibly happen, how likely it would happen, and what the consequences are if it does happen. The FRA involves several steps, including identifying the objectives of the assessment, the metrics for assessment, the hazards of concern and the potential fire scenarios, conducting frequency and consequence analyses on the scenarios of concern, and estimating the risk associated with the scenarios. The process is shown in Fig. A.2.

The use of risk assessment techniques will allow translation to risk management and risk-informed decision making. There are variety of methods to conduct risk assessment exercises, which are dependent on the level of detail needed and available data [3]. The assessment can be done in three levels, purely qualitative methods, semi-quantitative methods, and purely quantitative methods. A qualitative risk assessment analysis requires the least level of detail and the least data, where a quantitative risk assessment analysis requires the highest level of detail and more data. Types of analysis that can be done range from a simple unstructured method to a quantitative risk assessment (QRA) or probabilistic risk analysis (PRA). Listed are some other examples:

- HAZAN—Hazard Analysis
- Matrix Methods
- GOFA—Goal Oriented Failure Analysis
- HAZID—Hazard Identification
- HAZOP—Hazard and Operability Study
- Consequence Analysis (CA)
- Fault Tree Analysis (FTA)
- Event Tree Analysis (ETA)
- Cause-Consequence Analysis
- Failure Modes, Effects, and Criticality Analysis (FMECA).

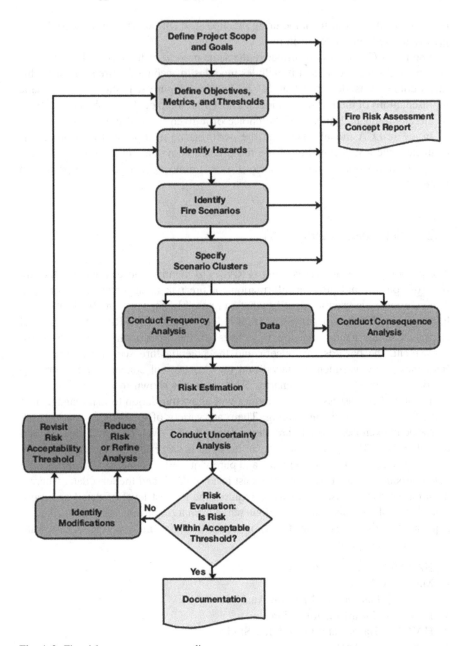

Fig. A.2 Fire risk assessment process diagram

The process of risk assessment can be conducted through a variety of methods depending on the problem being studied and the results that are desired. The process of assessing the risk of fire on the environment requires additional study because of

Fig. A.3 EPA Risk
Assessment Framework
(EPA 1998)

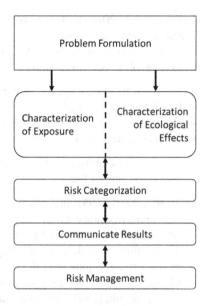

the gaps in some of our knowledge as well as the specificity and complexity of the issues involved. The framework shown in Fig. A.3 is used by Environmental Protection Agency (EPA) to conduct their ecological risk assessment.

A.2.1 Economic Assessment (EA)

A.2.1.1 Cost Benefit Analysis

The cost-benefit analysis (CBA) is a technique used to look at products, processes, or legislation to objectively quantify the costs involved and compares them with their benefits [3]. The cost-benefit analysis will not only account for the direct economic costs, but for the costs to the whole process, product, or legislation. Similarly to the LCA, the cost benefit analysis requires quantified data for items that have not traditionally been communicated with quantified data. An exploratory case study of economic assessment was conducted by the Seidman Research Institute when the City of Phoenix Fire Department saved a furniture manufacturer from a major fire. The economic impact of successful fire interventions at commercial establishments over a 12 months period was further analyzed.

The Economic Impact of successful commercial fire interventions—Phoenix Fire Department (Arizona State University 2012).

An Arizona specific version of the Regional Economic Models Inc. (REMI) model, created at the Seidman Research Institute, was used to estimate the economic impact of the City of Phoenix Fire Department's intervention at 42 fires, starting June 1, 2012 through May 31, 2013. REMI is a market-leading, dynamic forecasting and analysis tool that can help track the economic impact of a business at different moments in time. Seidman's method for estimating the economic impacts involves following steps:

- Prepare a baseline forecast for the state and county economy
- Develop policy scenario
- Compare the baseline and policy scenario forecasts
- Produce delta results.

As a result of impact analysis, the model was able to estimate some negative economic impacts of fires:

- Approximately 6951 total private non-farm jobs could have been lost in the State of Arizona over the course of one year if the City of Phoenix Fire Department had not successfully intervened at the forty-two commercial fires studied.
- If government and farm sector employment is included, the total impact could increase to 7446 jobs over the course of just one year in the State of Arizona.
- Maricopa County, as the host county, could suffer most of the estimated job losses, including 3023 full-time direct jobs.
- Gross state product could be lower by almost $650 million (2012 $) throughout the State of Arizona, and real disposable personal income by $295.6 million (2012 $), without the City of Phoenix Fire Department's successful interventions at these forty-two commercial fires.
- State tax revenues could also fall by over $35 million (2012 $) throughout the State of Arizona if the commercial fires had not been extinguished.
- The City of Phoenix Fire Department is therefore estimated to exert a significant impact on the local economy at both a state and county level over a twelve month time horizon, exclusively based on their successful commercial fire interventions.
- With successful commercial fire interventions accounting for only 3 % of the City of Phoenix Fire Department's annual workload, the study estimates are in all probability a conservative measure of the Fire Department's total annual economic impact for Maricopa County and the State of Arizona.

When a commercial business or organization is affected by fire, causing a temporary or permanent cessation of trade or potentially even relocation, this will also affect the host state or county's local economy. The potential impacts of fire damage include physical structure impairment, falls in sales output, or new production costs such as the purchase of replacement equipment and supplies. Those loss will affect economic variables such as employment, gross state product, disposable personal income and local/state tax revenues. For the purpose of this study, a commercial business or organization was assumed to be forced to temporarily or permanently close down due to fire when potential economic losses was estimated. No consideration was given to the potential construction impacts arising from unsuccessful interventions. Residential interventions, and other City of Phoenix Fire Department activities were also excluded from the analysis.

A.3 Related Regulatory Frameworks

ASTM E2921: Standard Practice for Minimum Criteria for Comparing Whole Building Life Cycle Assessments for Use with Building Codes and Rating Systems

EN 15978: Sustainability of construction works. Assessment of environmental performance of buildings. Calculation method

ISO 14040: Environmental management—Life cycle assessment—Principles and framework

ISO 14041: Environmental management—Life cycle assessment—Goal and scope definition and inventory analysis

ISO 14042: Environmental management—Life cycle assessment—Life cycle impact assessment

ISO 14043: Environmental management—Life cycle assessment—Life cycle interpretation

ISO 14044: Environmental management—Life cycle assessment—Requirements and guidelines

ISO 14045: Environmental management—Eco-efficiency assessment of product systems—Principles, requirements and guidelines

ISO 14047: Environmental management—Life cycle assessment—Illustrative examples on how to apply ISO 14044 to impact assessment situations

ISO 14071: Environmental management—Life cycle assessment—Critical review processes and reviewer competencies

ISO 16732-1: Fire safety engineering—Fire risk assessment—Part 1 General

ISO 16732-3: Fire safety engineering—Fire risk assessment—Part 3 Example of an industrial facility.

A.4 Important Parameters

When conducting LCA, RA or EA, each assessment requires very similar basic parameters. Here are the primary parameters to be considered [1]:

- Energy use
- Resource use
- Emissions (important parameter for fire)
- Waste.

More specific data are required depending on the type of analysis and the focus of the study. For example, CBA would require the information on total cost spent on a building, employee, transportation, products and the total cost of loess due to fire. More detailed and available the information, more accurate the analysis become. The relative effort that is required for the LCA is [3]. For example, Hamzi [2] shows steps to conduct Fire-LCA on storage tanks used in crude oil material production include:

- Type of complex where accidents occurred
- Type of tank contents
- Type of accident (fire, explosion, spill, etc.)
- Cause of Tank Accidents (lightening, maintenance, failure, etc.)
- Size of tank
- Fire emissions.

Here are some examples of specific parameters to be considered for focused assessments:

- Occupancy and construction types
- Travel time and distance to a fire incident
- Number and type of vehicles responded
- Percent of property burned/saved
- Suppression/mitigation media used
- Waste handling and recycling procedures
- Transportation generation.

A.5   Statistics

NFPA's Fire Analysis and Research division provides reports and statistics on the loss of life and property from fire events. NFPA estimates fires and fire losses by analyzing each data element in National Fire Incident Reporting System (NFIRS).

Here are some statistics that are available publicly:

- Fires in the U.S.
- Overall fire problem
- Fire loss in the U.S.
- Total cost of fire in the U.S.
- Trends and patterns of U.S. fire losses
- Number of fires by occupancy/property type
- Number of fires by type of fire
- Number of non-home structure, non-residential structure and structure fires
- Multiple-death fires
- Catastrophic multiple-death fires
- Home fires with 10 or more deaths
- Deadliest fires and explosions by property class
- Deadliest single building or complex fires and expositions in the U.S.
- Deadliest fires and explosions in U.S. history
- Deadliest fires and explosions in the world
- Large property loss
- Large-loss fires in the U.S.
- Largest fire losses in the U.S. history.

Fire causes:

- Appliances and equipment
- Cooking, heating, computer rooms and electronic equipment areas, office equipment, torches and burners, air conditioning, clothes dryers and washing machines, and kitchen equipment not including cooking equipment
- Arson and juvenile fire setting
- Playing with fires
- Intentional fire
- Candles
- Chemical and gases
- Spontaneous combustion or chemical reaction
- Flammable gas or combustible liquid
- Electrical
- Fireworks
- Holiday
- Christmas tree, holiday lights, and decorations
- Household products
- Mattresses, bedding, and upholstered furniture
- Lightning fires and lighting strikes
- Smoking materials.

Fire Services

- Administration
- Needs assessment
- Fire departments in Canada
- Fire service performance measures
- U.S. fire department profile, firefighters and fire departments
- Number of U.S. firefighters
- Firefighting occupations by women and race
- Fatalities and injuries
- Firefighter fatalities and injuries in the U.S.
- Firefighter deaths and injuries by cause/nature of injury, by type of duty, and by year
- Patterns of firefighter fireground injuries
- U.S. firefighter deaths related to training (2009–2010)
- U.S. volunteer firefighter injuries
- Incidents with 8 or more firefighter deaths
- Deadliest wildland firefighter fatality incidents
- Fire department calls
- False alarm activities in the U.S.
- Unwanted fire alarms
- Fire department calls by type and year.

Fire Safety Equipment

- Smoke alarms in the U.S. home fires
- U.S. experience with sprinklers
- Non-water based automatic fire extinguishing equipment.

Fires by property types

- Assemblies
- Eating and drinking establishments
- Religious and funeral properties
- Business and mercantile
- Service stations
- Stores and other mercantile properties
- Office properties
- Educational properties
- Largest educational structure fire losses
- School fires with 10 or more deaths
- Health care facilities
- High-rise building fires
- Industrial and manufacturing facilities
- Petroleum refineries and natural gas plants
- Prisons and jails
- Residential
- Dormitories, fraternities, sororities and barracks
- Residential properties under construction or undergoing major renovation
- Board and care facilities
- Hotel and morel structure fires
- Apartment, home, one- and two- family home, and residential structure fires by year
- Storage
- Barns
- Outside storage tanks
- LP-gas bulk storage
- U.S. structure fires in barns and warehouses
- Vacant buildings.

Outdoor Fires

- Brush, grass and forest fires
- Largest loss in wildland fires.

Vehicles

- Vehicle fire trends and patterns
- Industrial loaders and forklifts
- Buses
- Highway vehicle fires by year.

The USFA collects data from variety of sources to provide information on fire problems in the U.S. USFA references NFPA for some statistics. Here are some statistics that are available publicly:

- Trends in fires, deaths, injuries and dollar loss
- Causes of fires
- Residential
- Non-residential
- Vehicle
- Outside fires
- Fires in other places
- Where fire occurs
- Property types
- Impact of fire
- Fire death and injury
- Gender and Race
- Age
- Firefighters and fire departments
- Fire loss calculation.

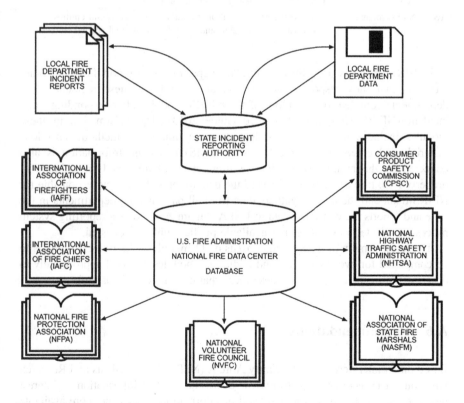

Fig. A.4 Incident reporting system

Table A.1 Types of modules used to collect information into NFIRS

ID	Module	Description
1	Basic	General information. This module is required
2	Fire	Used to report fire incident information
3	Structure fire	Used to capture structure fire information
4	Civilian fire casualty	Used to report fire-related civilian injuries or deaths
5	Fire service casualty	Used whenever there is a fire service casualty
6	EMS	Used only if the fire department provides EMS
7	Hazardous material	Used when hazardous material resources are dispatched, released or spills exceeding 55 gallons
8	Wildland fire	Used to capture data about the number of acres burned, type of materials involved, conditions that contributed to the spread of fires, and resources needed to control or extinguish fires
9	Apparatus or resources	Used to manager and track apparatus and resources used at an incident
10	Personnel	Used to manager and track personnel and resources used at an incident
11	Arson	Used for intentionally set fire information
1S	Supplemental form	Used to report information on additional persons and entities involved in the incident and to collect additional special studies

NFIRS website is used by the U.S. fire departments to report fires and other incidents that they respond, and to maintain records of these incidents. Fire departments complete one or more of the NFIRS modules after responding to an incident (NFIRS Complete Reference Guide 2013). The information in these modules includes the kind of incident, location, mitigation method, hazardous material, and emergency medical service (EMS) incidents. Information is also collected on the number of civilian or firefighter casualties and an estimate of property loss. Fire departments forward the data to the state agency responsible for NFIRS data, and the agency gathers data from all participating departments in the state and reports the compiled data to USFA. Figure A.4 shows a reporting system process. Collected raw data is not available publicly unless requested.

The latest version of the system consists of 11 modules. The Basic Module is to be completed for every incident, and additional modules are used appropriately to describe the incident. Table A.1 describes what each module is used for.

A.6   Recommendations

Based on an interview with Ms. Marty Ahrens (NFPA, Fire Analysis and Research Division), it is recommended to narrow down the level of information of interest when collecting statistics. National level of information tends to be more available with less details, while local level of information might be more difficult to look for and provide more detailed data. At this point, we have decided to look into warehouse fire events that are (1) representative, (2) well known and well documented, and (3) the case where water was used to extinguish the fire. On top of this, we should consider deciding whether a case should include fire detection, automatic sprinkler system, a cold storage (warehouse with or without cold storage is a separate category in NFIRS database), use of chemical suppression, and presence of large loss.

Available statistics summarized in this book suggest that it seems to be difficult to conduct CBA with the level of detail we are provided. Not all cost associated with every single component of the process when fire happens can be found in one place. Information is collected by fire departments, not by fire events, which makes an incident analysis difficult. For example, a cost of fire brigade intervention is extremely difficult to estimate because functions such as paid/volunteer firefighters, resources used, location and training of each fire department involved in the event need to be considered.

Ms. Marty Ahrens also mentioned that International Association of Fire Fighters (IAFF) is compiling detailed national fire database within a year or so. This might something to look into for the future project.

References

1. Andersson, P., Simonson, M., Tullin, C., Stripple, H., Sundqvist, J., Paloposki, T.: Fire-LCA guidelines. SP Fire Technology. Retrieved from http://www.sp.se/en/index/services/firelca/sidor/default.aspx (2004)
2. Hamzi, R., Londiche, H., Bourmada, N.: Fire-LCA model for environmental decision-making. Chem. Eng. Res. Des. **86**(10), 1161–1166. doi:10.1016/j.cherd.2008.05.004. Retrieved from http://www.sciencedirect.com/science/article/pii/S0263876208001433 (2008)
3. Martin, D., Tomida, M., Meacham, B.: The environmental impact of fire. NFPA-Fire Protection Research Foundation. National Fire Protection Association, 1 May 2015. Retrieved from http://www.nfpa.org/research/fire-protection-research-foundation/projects-reports-and-proceedings/other-research-topics/the-environmental-impact-of-fire (2015)
4. Meacham, J.B.: An overview of approaches and resources for building fire risk assessment. Fire Protection Engineering. Society of Fire Protection Engineering, 1 July 2013. Retrieved from http://magazine.sfpe.org/content/overview-approaches-and-resources-building-fire-risk-assessment.
5. National Fire Incident Reporting System (NFIRS): Complete reference guide. National Fire Incident Reporting System (NFIRS). Retrieved from https://www.nfirs.fema.gov/
6. Reports and Statistics: National Fire Protection Association. Retrieved from http://www.nfpa.org/research/reports-and-statistics

7. Simonson, M., Andersson, P., Tullin, C., Stripple, H., Sundqvist, J., Paloposki, T.: Fire-LCA Guidelines. SP Fire Technology (2005)
8. The Economic Impact of Successful Commercial Fire Interventions: Arizona State University. Retrieved from http://media2.abc15.com/html/pdf/ASU%20study.pdf (2012)
9. U.S. Fire Statistics: U.S. Fire Administration. Retrieved from http://www.usfa.fema.gov/data/statistics/

Appendix B
Historic Warehouse Fires as Potential Case Studies

by Mai Tomida and Brian Meacham
Worcester Polytechnic Institute
Department of Fire Protection Engineering
50 Prescott Street
Worcester, MA 01609

This book summarizes the selection process of warehouse case studies and recorded data for each study that can be used for an environmental/economic assessment tool. The tool is being developed based on available statistics of fire events to estimate the impact of fire. This investigation is currently focused on U.S. data organized by National Fire Protection Association (NFPA). Documents that are highlighted in this book were collected from NFPA and U.S. Fire Administration (USFA) websites.

B.1 Methodology

Facilities classified as a warehouse were selected as a starting structure of study. Following three criteria were used to collect documents: (1) well documented, (2) representative and (3) water was used to extinguish the fire. NFPA Fire Investigation Reports were used as a main resource. The areas documented during the investigations include details of fire ignition, growth, and development; contributions of building construction, interior finish and furnishings; fire detection and suppression scenarios; performance of structures exposed to the fire; smoke movement and control; human reaction (response) and evacuation; firefighting and rescue; fire propagation as a function of human reaction time; and the extent of life loss, injury and property damage.

Another recourse used to collect fire event documents under the above criteria was USFA. The USFA develops reports on selected major fires throughout the country, where fires usually involve multiple deaths or a large loss of property. The primary criterion for deciding to do a report is whether it will result in significant "lessons learned." Some lessons bring to light new knowledge about fire, such as

© Fire Protection Research Foundation 2016
F. Amon et al., *Development of an Environmental and Economic Assessment Tool (Enveco Tool) for Fire Events*, SpringerBriefs in Fire, DOI 10.1007/978-1-4939-6559-5

the effect of building construction or contents and human behavior in fire. Some special reports are developed to discuss events, drills, or new technologies that are of interest to the fire service.

Using these recourses, case studies focused on warehouse fire were collected and further looked into the type of available data. Since the necessary parameters for the assessment tool have not finalized at the moment, cases with more available statistics were selected.

B.2   Warehouse Fires Literature Review

Some of the warehouse fires documented by NFPA and USFA are summarized in this section. Events were selected based on the available information, such as occupancy classification, construction type, use of sprinkler system, fire mitigation method, and so on. As requested, documents were collected under following criteria: (1) well documented, (2) representative and (3) water was used to extinguish the fire.

B.2.1   Cold Storage Warehouse Fire (Madison, WI)

Overview
May 3, 1991

The estimated loss was more than $100 million for a fire of undetermined origin that destroyed two cold storage warehouses in Madison, Wisconsin, on May 3, 1991. The warehouses were part of a five building complex, where food was stored including 13 million pounds of butter and 15 ½ million pounds of cheese. All of the buildings were constructed with large metal structures, and several were equipped with automatic sprinkler systems.

The warehouse complex consisted of five cold storage buildings, which were built with light, combustible Type II (000) construction. The exterior walls of the building's freezer areas were no-loadbearing, with foam insulation between metal sheathing. A layer of foam insulation over the roof's metal deck was covered with tar and gravel. A single ammonia refrigeration system was used to protect the facility's coolers and freezers.

Available Information:

- The estimated loss: $100 million
- Construction type: Type II (000)
- Stored commodities: butter and cheese
- Exterior finish: no-loadbearing, foam insulation between metal sheathing
- Roof covering material: tar and gravel
- Are of the fire origin: 170 ft wide, 365 ft long, 55 ft high

- Freezer/cooler dimension, interior finish, temperature, commodities and piping
- fire alarm detection time, number of crew, and total time on extinguishing effort
- Damage results
- The cause of fire analysis
- sprinkler system design
- size and location of the building.

B.2.2 Food Processing Plant Fire (Yuma, AZ)

Overview
November 12, 1992

An accidental fire destroyed most of the Dole Fresh Vegetables plant in Yuma, Arizona, resulted in a loss estimated as $16 million. The building was a noncombustible structure with light-gauge metal exterior walls and roof. Polyurethane foam insulation was sprayed over the interior surfaces. The building was equipped with automatic sprinkler systems.

A welder was installing process equipment during the construction of an additional facility, and may have accidentally ignited combustible materials including the sprayed-on foam insulation inside a wall assembly. The fire spread in a combustible concealed space between the wood-framed interior walls and the metal exterior walls, where sprinklers had not been installed. As a result, the sprinkler systems that operated were not able to control the fire spreading within the walls.

Available Information:

- The estimated loss: $16 million
- Construction type: non combustible
- Interior finish: polyurethane foam spray insulation
- Use of automatic sprinkler system
- Area of the facility
- Layout of the facility
- Roof/ceiling assembly
- Sprinkler system: hydraulically calculated 6 automatic sprinklers, piping layout
- Facility considered as Ordinary Hazard Group II based on NFPA 13
- Fire alarm detection time
- Number of firefighters
- Extinguishing time
- Damage results
- The cause of fire analysis
- Construction process, interior insulation
- Fire growth timeline.

B.2.3 Furniture Manufacturing Facility Dust Explosion (Lenoir, NC)

Overview
November 20, 1994

A series of explosions occurred at a furniture manufacturing facility in Lenoir, North Carolina, including two fatalities and four injuries as a result. The incident took place in the particleboard manufacturing portion of the plant. Four explosions followed the initiating event, caused by the dust in the facility. The total damage to the facility was estimated to be 139,000 ft^2, and the production was interrupted for over nice months.
Available Information:

- Number of employees: approximately 1500 people
- Area of the facility
- Occupancy classification: Extra Hazard Occupancy Group 1 (MFPA 13, 1994), and Special Industrial Ordinary Hazard (NFPA 101, 1994)
- Construction type: Type II (000) (NFPA 220, 1992)
- List of NFPA codes applied to the facility
- Exterior finish: lightweight corrugated sheet metal or composite panels
- Fire protection systems: sprinkler, fire alarm and special suppression systems throughout the complex, piping, private water supply, and fire pump. The area involved in the incident was protected by 3 sprinklers
- Fire fighting strategy: rescue, total time that the team was present
- Damage results
- The cause of fire analysis: thermal oil transfer unit room or the milling and drying area, broken mains of the sprinkler system.

B.2.4 Storage Warehouse Fire (Phoenix, AZ)

Overview
August 2nd, 2000

A fire was discovered in a multi-tenanted warehouse in Phoenix, AZ. This fire destroyed the 85,000 ft^2 warehouse by the time it was extinguished. The damage to the property and the commodities stored from this fire was estimated to be more than $100.
Available Information:
Occupancy type: Storage (NFPA101, 2000)

- Layout of the facility
- Total area of the facility

- Roof covering: Plywood deck supported by wood laminate rafters on steel columns, multiplayers of mopped asphalt on the plywood decking
- Fire protection systems: automatic sprinkler system, design details
- Stored commodities: Lawn and garden care products, pool care products, landscaping materials, artificial fireplace logs, hand tools, lawn mower accessories, wood planters and clay pots, etc.
- Local fire department
- Weather
- Fire fighting strategy: fire alarm detection time, number of crew, time line of fire extinguishing effort, layout of operation
- Damage results
- List of oxidizers
- List of NFPA codes applied to the facility.

B.2.5 Warehouse Fire (New Orleans, LI)

Overview
March 21, 1996

Two fires occurred in an occupied, operational general merchandise warehouse in New Orleans, Louisiana. The warehouse area of the building measured 930,020 ft^2. This warehouse contained a combination of high racks measuring 65 ft high and low racks measuring 21–30 ft high. High racks were all equipped with in-rack sprinklers, where not all of the low racks were equipped. The main warehouse area housed a variety of conveyor systems and a mechanized retrieval system. There were 15 employees in the building at the time of fire. The sprinkler systems were not able to control the fire for the second time, and the warehouse and distribution area were destroyed.

Available Information:

- Occupancy type: Storage, Ordinary Hazard
- Construction type: Type II (000), (NFPA 220, 1995)
- Total area of the facility: the warehouse part is 930,020 ft^2
- Layout of the facility
- Stored commodities: wicker baskets and furniture, rugs, polyfill pillows, cardboard boxes, comforters, towels, stacked plastic chairs and plastic bags...etc
- Fire protection systems: warehouse was protected by 30 overhead sprinkler systems and 17 in-rack sprinkler systems, no fire separation barriers
- Fire fighting strategy: fire alarm detection time, number of crew, time line of fire extinguishing effort
- Damage results
- The cause of fire analysis: sprinkler system performance, lack of fire separation.
- List of NFPA codes applied to the facility.

B.2.6   Abandoned Cold Storage Warehouse Multi-firefighter Fatality Fire (Worcester, MA)

Overview

A smoke coming from the roof of the Worcester Cold Storage and Warehouse Co. was reporter on Friday, December 3, 1999 in Worcester, MA. The original building was constructed in 1906, contained 43,000 ft^2 and 6 stories above grade. The building was known to be abandoned for over 10 years. The first alarm assignment brought 30 firefighters and officers and 7 pieces of apparatus to the scene. The second provided an additional 12 men and 3 trucks as well as a Deputy Chief. Firefighters encountered a light smoke condition throughout the warehouse, and crews found a large fire in the former office area of the second floor. An aggressive interior attack was started within the second floor and ventilation was conducted on the roof. There were no windows or other openings in the warehousing space above the second floor.

An extensive search of two homeless people was conducted by Worcester Fire crews through the third and fourth alarms. Suppression efforts continued to be ineffective against huge volumes of petroleum based materials. When the evacuation order was given one hour and forty-five minutes into the event, five firefighters and one officer were missing. None survived.

According to NFPA records, this is the first loss of six firefighters in a structure fire where neither building collapse nor an explosion was a contributing factor to the fatalities.

Available Information:

- Occupancy type and building use: Cold storage
- Construction type: Type II (000)
- Building structure details: area, interior/exterior finish, roof/ceiling assembly
- Fire protection: non-functional suppression system working at the time of fire
- Involved commodities: carpet, mild crates, candle
- Firefighting strategy: number of crew, time line of fire extinguishing effort
- Site/floor plan
- The cause of fire analysis and lessons learned.

B.2.7   Sherwin-Williams Paint Warehouse Fire (Dayton, OH)

Overview

The fire started on May 27, 1987, and completely destroyed the Sherwin-Williams Paint Warehouse in Dayton, OH. The dollar loss was estimated to be $32 million, one employee was seriously injured and one firefighter sprained his leg. The noncombustible, sprinklered warehouse contained over 1.5 million gallons of paints and other products. It was located over the aquifer from which wells provided the water supply for about one-third of the area's 400,000 people. Uncontained water and chemical run-off from firefighting could have contaminated this water supply and caused a greater loss than the fire itself, as occurred in Switzerland after the Sandoz Chemical Warehouse fire in 1986 contaminated the Rhine. The Dayton, Ohio Fire Department avoided a double disaster by not attempting to extinguish a massive fire in a paint warehouse.

Available Information:

- Occupancy type and building use: one-story warehouse attached to an office building
- Building structure details: area, interior/exterior finish, roof/ceiling assembly
- Fire protection: sprinkler system, diesel fire pump (2500 gpm) supplied by 12 inch line connected to 16 inch public water main
- Stored commodities: over 1.5 gallons of paints and related flammable liquids
- Number of employees working during the event: 30
- Firefighting strategy: number of crew, time line of fire extinguishing effort
- Site/floor plan
- The cause of fire analysis and lessons learned
- NFIRS incident reports.

B.2.8 Sandoz Chemical Plant Fire (Basel, Switzerland)

Overview

Thirty tons of toxic material washed into the Rhine River with water firefighters used to fight a warehouse blaze at a riverside Sandoz Chemical Plant and Storage Facility near Basel, Switzerland on November 1, 1986. By the time the chemicals, mostly pesticides, had traveled 500 miles down the winding scenic river, half a million fish were dead, several municipal water supplies were contaminated, and the Rhine's ecosystem was badly damaged with virtually all marine life and a large proportion of microorganisms wiped out. The approximately 25 mile long chemical slick drifted slowly downstream from the Swiss border to the North Sea. It contained about 30 tons of insecticides, herbicides, and mercury containing pesticides, and threatened the North Sea's winter cod harvest. The warehouse was built in 1967 and was approximately 295 ft long by 82 ft wide. This is a very well known event where results of the incident caused both environmental and economic impact negatively on Switzerland and surrounding country afterwards.

Available Information:

- Occupancy type: warehouse
- Building structure details: area, interior/exterior finish, roof/ceiling assembly
- Fire protection: no sprinkler system,

- Stored commodities: 1250 tons of chemicals in barrels, mainly flammable including pesticides, fungicides and herbicides
- Site/floor plan
- Breakdown of the cost associated with the incident
- Damage results.

B.3 Recommendation

For selecting and using the best case study for the environmental/economic assessment tool, I recommend finalizing the parameters essential to the tool and clearly identify a set of requirements for the model. Cases introduced in this book were selected based on the available amount of information. There was commonly available information within NFPA reports, such as occupancy classification and construction type of the building, area of the incident, stored commodities, fire protection systems (sprinklers and alarms), local fire department information, damage results of the fire and so on. NFPA reports also seemed to be more statistically well organized than USFA reports. This might be because NFPA seems to investigate the fire systematically, and USFA seems to focus on investigating the cause and lessons learned. This might be interesting to consider when designing a tool.

There were no CO_2 emissions reported in most of the documents reviewed. This could be because an environmental issue was not part of the study, or the main concern back then, considering that these reviewed reports are mostly from 1990 to 2000. When designing a tool, certain assumptions might be required to make especially when estimating the amount of environmental pollutants.

To further narrow down the case studies selected, identification of parameters for the tool is required. Once requirements for the model are finalized, selected documents can be revisited to determine the appropriate case study for testing the model. It is also possible to prioritize what information is needed the most after defining important parameters.

References

1. Anderson, J.: Abandoned cold storage warehouse multi-firefighter fatality fire. U.S. Fire Administration, 1 Dec 1999. https://www.usfa.fema.gov/downloads/pdf/publications/tr-134. pdf?utm_source=website&utm_medium=pubsapp&utm_content= AbandonedColdStorageWarehouseMulti-FirefighterFatalityFire(Worcester,MA-December1999)&utm_campaign=TDL
2. Comeau, E., Puchovsky, M.: Warehouse fire New Orleans, Louisiana. Fire Investigations (Print)
3. Comeau, E.: Furniture manufacturing facility dust explosion, Lenoir, NC. Fire Investigations (1995) (Print)

4. Copeland, T., Schaenman, P.: Sherwin-Williams paint warehouse fire. U.S. Fire Administration. https://www.usfa.fema.gov/downloads/pdf/publications/tr-009.pdf
5. Duval, R.: Storage warehouse, Phoenix, AZ. Fire Investigations (2002) (Print)
6. Isner, M.: Food processing plant, Yuma, Arizona. Fire Investigations (Print)
7. Isner, M.: Cold storage warehouse, Madison, WI. Fire Investigations (1991) (Print)
8. Schwabach, A.: The Sandoz spill: the failure of international law to protect the rhine from pollution. Ecology Law Quarterly. University of California Berkeley Law. http://scholarship. law.berkeley.edu/cgi/viewcontent.cgi?article=1355&context=elq
9. The Rhine Polluted by Pesticides: French Ministry of the Environment. http://www.aria. developpement-durable.gouv.fr/wp-content/files_mf/FD_5187_schwizerhalle_1986_ang.pdf

Appendix C
Statistical Decision Support for Possible Future Expansion of the Enveco Tool

by Mai Tomida and Brian Meacham
Worcester Polytechnic Institute
Department of Fire Protection Engineering
50 Prescott Street
Worcester, MA 01609

Available statistics were revisited to examine the possible expansion of the environmental/economic assessment tool to include other structures besides warehouses. Statistics from National Fire Protection Association (NFPA), U.S. Fire Administration (USFA) and U.S. Energy Information Administration (USEIA) were used to prioritize this expansion. Important criteria considered in this process was: (1) impact/relevance to the society and (2) availability of statistics.

C.1 NFPA

C.1.1 CStructure Fires by Occupancy (April 2013)

This NFPA report provides a collection of tables of the estimated average number of structure fires, associated civilian deaths and injuries, and direct property damage in terms of occupancy types. The analysis is based on fires per year incidents reported to local fire departments during 2007–2011. The estimates were derived from Version 5.0 of the USFA's National Fire Incident Reporting System (NFIRS) and NFPA's annual fire department experience survey. Figure C.1 shows a general detail of the number of fires in each occupancy types. It is shown that Residential occupancy takes up 79 % of the fire events, 96 % of civilian deaths, 90 % of civilian injuries, and 71 % of direct property damage. It is clear that this is the most affected occupancy type. Business/Mercantile occupancy seems to have the second effected area with 4 % fire events, 1 % civilian deaths, 2 % civilian injuries, and 8 % direct property damage. The full details of these occupancies are shown in Figs. C.2 and C.3.

© Fire Protection Research Foundation 2016

F. Amon et al., *Development of an Environmental and Economic Assessment Tool (Enveco Tool) for Fire Events*, SpringerBriefs in Fire, DOI 10.1007/978-1-4939-6559-5

Occupancy	Fires		Civilian Deaths		Civilian Injuries		Direct Property Damage (in Millions)	
1 - Assembly	14,650	(3%)	5	(0%)	173	(1%)	$425	(4%)
2 - Educational	5,690	(1%)	1	(0%)	85	(1%)	$92	(1%)
3 - Health Care, Detention & Correction	6,820	(1%)	5	(0%)	172	(1%)	$57	(1%)
4 - Residential	392,800	(79%)	2,696	(96%)	13,754	(90%)	$7,605	(71%)
5 - Mercantile, Business	17,740	(4%)	22	(1%)	325	(2%)	$810	(8%)
6 - Industrial, Utility, Defense, Agriculture, Mining	2,860	(1%)	4	(0%)	51	(0%)	$232	(2%)
7 - Manufacturing, processing	5,300	(1%)	7	(0%)	170	(1%)	$593	(6%)
8 - Storage	21,770	(4%)	30	(1%)	283	(2%)	$614	(6%)
9 - Outside or special property	22,330	(4%)	15	(1%)	113	(1%)	$81	(1%)
Unclassified, unreported, and unknown	8,540	(25)	13	(0%)	94	(1%)	$142	(1%)
Totals	498,500	(100%)	2,798	(100%)	15,221	(100%)	$10,650	(100%)

Fig. C.1 Structure fires by occupancy, general detail

Occupancy	Fires		Civilian Deaths		Civilian Injuries		Direct Property Damage (in Millions)	
Residential								
419-One-or two-family homes	260,180	(52%)	2,165	(77%)	8,931	(59%)	$5,959	(56%)
429-Apartment of multi-family home	106,380	(21%)	410	(15%)	4,276	(28%)	$1,248	(12%)
439 –Includes residential hotels and shelters Boarding rooming house	2,850	(1%)	18	(1%)	87	(1%)	$33	(0%)
449-Hotel or motel	3,610	(0%)	11	(1%)	139	(1%)	$125	(1%)
459 - Residential board and care	1,860	(0%)	6	(0%)	50	(0%)	$9	(0%)
460 - Dormitory type residence, other	3,040	(1%)	2	(0%)	25	(0%)	$5	(0%)
462 - Sorority house, fraternity house	200	(0%)	0	(0%)	2	(0%)	$3	(0%)
464 - Barracks, dormitory	570	(0%)	0	(0%)	3	(0%)	$2	(0%)
400 - Residential, other	14,130	(3%)	85	(3%)	242	(2%)	$221	(2%)

Fig. C.2 Residential fires by occupancy, full detail

Figure C.2 suggests that one- or two-family homes is the most affected category within the residential occupancy, followed by apartment or multi-family home and so on.

Figure C.3 suggests that food and beverage sales, grocery store is the most affected category within the business/mercantile occupancy, followed by business office and so on.

C.1.2 Fire Loss in the U.S. During 2014 (September 2015)

This report provides losses that were cause by fires in 2014, U.S. fire departments responded to an estimated 1,298,000 fires. These fires resulted in 3275 civilian fire fatalities, 15,775 civilian fire injuries and an estimated $11.6 billion in direct

Occupancy	Fires		Civilian Deaths		Civilian Injuries		Direct Property Damage (in Millions)	
Mercantile, business								
511 - Convenience store	820	(0%)	1	(0%)	11	(0%)	$26	(0%)
519 - Food and beverage sales, grocery store	3,080	(1%)	1	(0%)	51	(0%)	$85	(1%)
529 - Textile, wearing apparel sales	300	(0%)	1	(0%)	7	(0%)	$20	(0%)
539 - Household goods, sales, repairs	310	(0%)	0	(0%)	5	(0%)	$23	(0%)
549 - Specialty shop	1,420	(0%)	4	(0%)	20	(0%)	$99	(1%)
557 - Personal service, including barber & beauty shops	520	(0%)	1	(0%)	7	(0%)	$19	(0%)
559 - Recreational, hobby, home repair sales, pet store	190	(0%)	1	(0%)	4	(0%)	$13	(0%)
564 - Laundry, dry cleaning	1,050	(0%)	1	(0%)	20	(0%)	$22	(0%)
569 - Professional supplies, services	480	(0%)	0	(0%)	6	(0%)	$30	(0%)
571 - Service station, gas station	510	(0%)	0	(0%)	12	(0%)	$19	(0%)
579 - Motor vehicle or boat sales, services, repair	1,540	(0%)	5	(0%)	69	(0%)	$122	(1%)
580 - General retail, other	710	(0%)	1	(0%)	10	(0%)	$56	(1%)
581 - Department or discount store	430	(0%)	0	(0%)	6	(0%)	$19	(0%)
592 - Bank	270	(0%)	0	(0%)	2	(0%)	$7	(0%)
593 - Office: veterinary or research	90	(0%)	0	(0%)	4	(0%)	$6	(0%)
596 - Post office or mailing firms	80	(0%)	0	(0%)	0	(0%)	$1	(0%)
599 - Business office	2,890	(1%)	3	(0%)	38	(0%)	$98	(1%)
500 - Mercantile, business, other	3,050	(1%)	2	(0%)	52	(0%)	$145	(1%)

Fig. C.3 Business/mercantile fires by occupancy, full detail

property loss. Home fires caused 2745 (84 %) of the civilian fire deaths. Figures C.4 and C.5 show estimates of structure fires and property losses, and civilian fire deaths and injuries in terms of the property use respectively.

Figure C.4 suggests that residential occupancy, specifically one- and two-family homes, has the highest number of fires and the cost estimated, followed by storage in structures.

Figure C.5 suggests that residential occupancy, specifically one- and two-family homes, has the highest number of civilian deaths injuries estimated, followed by highway vehicles.

C.1.3 Trends and Patterns of U.S. Fire Losses in 2013 (March 2015)

This report provides an overview of historical perspective on the results based on NFPA's annual fire department experience survey published in the annual report, "Fire Loss in the United States". In 2013, structure fires accounted for 39 % of reported fires, with home structure fires representing 30 % of the total (Fig. C.6).

Property Use	Structure Fires		Property Loss[1]	
	Estimate	Percent Change from 2013	Estimate	Percent Change from 2013
Public Assembly	14,000	+12.0	$429,000,000	+16.3*
Educational	5,000	-9.1	$59,000,000	-10.6
Institutional	6,500	+8.3	$40,000,000	-4.8
Residential (Total)	386,500	+0.1	$6,992,000,000	+0.3
One- and Two-Family Homes[2]	273,500	+0.7	$5,844,000,000	+3.9
Apartments	94,000	-4.1	$982,000,000	-15.8*
Other Residential[3]	19,000	+8.6	$166,000,000	-6.2
Stores and Offices	17,500	-2.8	$708,000,000	+15.9
Industry, Utility, Defense[4]	10,000	+17.6	$626,000,000	-1.7
Storage in Structures	27,500	5.8	$781,000,000	+12.9
Special Structures	27,000	+12.3	$211,000,000	+50.7**
Total	494,000	+1.3	$9,846,000,000	+3.4

The estimates are based on data reported to the NFPA by fire departments that responded to the 2014 National Fire Experience Survey.

[1] This includes overall direct property loss to contents, structure, a vehicle, machinery, vegetation or anything else involved in a fire. It does not include indirect losses, e.g., business interruption or temporary shelter costs. No adjustment was made for inflation in the year-to-year comparison.

[2] This includes manufactured homes.

[3] Includes hotels and motels, college dormitories, boarding houses, etc.

[4] Incidents handled only by private fire brigades or fixed suppression systems are not included in the figures shown here.

*Change was statistically significant to the 0.05 level

**Change was statistically significant at the .01 level.

Fig. C.4 Estimates of 2014 structure fires and property loss by property use

Home structure fires caused 85 % of all civilian fire deaths, 77 % of civilian fire injuries, and 59 % of total direct property damage (Figs. C.7, C.8, and C.9).

Although the trends indicated that the total reported fires have been declining over the past 15 years due to a drop in the number of vehicle fires, outside and unclassified fires, the decline in structure fires has been much smaller.

| Property Use | Civilian Deaths | | | Civilian Injuries | | |
	Estimate	Percent Change From 2013	Percent of all Civilian Deaths	Estimate	Percent Change From 2013	Percent of all Civilian Injuries
Residential (total)	2,795	+0.4	85.3	12,175	-3.2	77.2
One- and Two-Family Homes[1]	2,345	-3.5	71.6	8,025	-3.3	50.9
Apartments	400	+23.1	12.2	3,800	-2.6	24.1
Other Residential[2]	50	66.7	1.5	350	-6.7	2.2
Non-Residential Structures[3]	65	-7.1	2.0	1,250	-16.7	7.9
Highway Vehicles	310	+3.3	9.5	1,275	+37.8	8.1
Other Vehicles[4]	35	+75.0	1.1	175	+40.0	1.1
All Other[5]	70	+7.7	2.1	900	+12.5	5.7
Total	3,275	+1.1		15,775	-0.9	

The estimates are based on data reported to the NFPA by fire departments that responded to the 2014 National Fire Experience Survey.

Note all of the changes were not statisically significant; considerable year-to-year fluctuation is to be expected for many of these totals because of their small size.

[1] This includes manufactured homes.

[2] Includes hotels and motels, college dormitories, boarding houses, etc.

[3] This includes public assembly, educational, institutional, store and office, industry, utility, storage, and special structure properties.

[4] This includes trains, boats, ships, farm vehicles and construction vehicles.

[5] This includes outside properties with value, as well as brush, rubbish, and other outside locations.

Fig. C.5 Estimates of 2014 civilian fire deaths and injuries by property use

Fig. C.6 Reported fire incidents by major property class (2013)

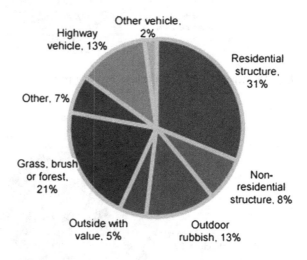

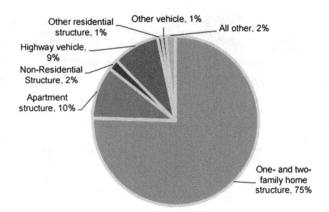

Fig. C.7 Civilian fire deaths by major property class (2013)

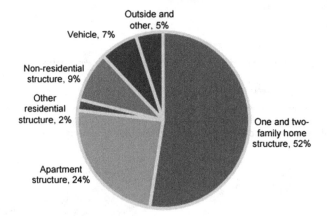

Fig. C.8 Reported civilian fire injuries by major property class (2013)

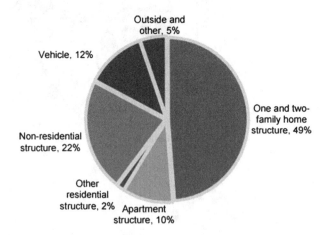

Fig. C.9 Direct property damage by major property class (2013)

Fires and Fire Losses by General Property Type (2011) *Totals may not add up to 100 percent due to rounding.

Fires by General Property Type (2011)			Fires Deaths by General Property Type (2011)			Fires Injuries by General Property Type (2011)			Fires Dollar Loss by General Property Type (2011)		
General Property Type	Percent of Fires		General Property Type	Percent of Fire Deaths		General Property Type	Percent of Fire Injuries		General Property Type	Percent of Fire Dollar Loss	
Residential	29.1%		Residential	75.7%		Residential	79.1%		Residential	52.2%	
Nonresidential	8.0%		Nonresidential	4.7%		Nonresidential	7.8%		Nonresidential	24.8%	
Vehicle	13.3%		Vehicle	14.6%		Vehicle	5.8%		Vehicle	10.2%	
Outside	44.0%		Outside	2.3%		Outside	3.2%		Outside	7.8%	
Other	5.7%		Other	2.7%		Other	4.2%		Other	5.1%	
Total	100.0%		Total	100.0%		Total	100.0%		Total	100.0%	
Source: National Fire Incident Reporting System			Source: National Fire Incident Reporting System			Source: National Fire Incident Reporting System			Source: National Fire Incident Reporting System		

Fire Casualties and Dollar Loss per Fire by General Property Type (2011)

Fire Deaths per 1,000 Fires (2011)			Fire Injuries per 1,000 Fires (2011)			Fire Dollar Loss per Fire (2011)		
General Property Type	Fire Deaths per 1,000 Fires		General Property Type	Fire Injuries per 1,000 Fires		General Property Type	Fire Dollar Loss per Fires (in $ thousands)	
Residential	5.3		Residential	28.8		Residential	14.8	
Nonresidential	1.2		Nonresidential	10.3		Nonresidential	25.7	
Vehicle	2.2		Vehicle	4.6		Vehicle	6.3	
Outside	0.1		Outside	0.8		Outside	1.5	
Other	1.0		Other	7.8		Other	7.4	
Source: National Fire Incident Reporting System			Source: National Fire Incident Reporting System			Source: National Fire Incident Reporting System		

Fig. C.10 Fires and fire losses by general property type (2011)

C.2 USFA

USFA's website states that "'Residential' is the leading property type for fire deaths (75.7 %), fire injuries (79.1 %) and fire dollar loss (52.2 %)". This can be seen on Fig. C.10 where fire statistics from 2011 are organized in terms of the property type.

There seem to be more reports available from USFA on residential fire events compared to warehouse fires. A quick search showed about 30 reports on residential fires, which is six times more than that of warehouse fires. A literature review on this can be done on this area if necessary.

C.3 USEIA

Variety of building data is available from the USEIA Website. Consumption surveys are conducted on residential, commercial building, manufacturing and vehicle areas to collect a wide range of information. All of the data is available freely on the website at: http://www.eia.gov/consumption/data.cfm#rec.

C.3.1 Residential Energy Consumption Survey (RECS)

The Residential Energy Consumption Survey (RECS) targets a nationally representative sample of housing units. Energy characteristics, usage patterns and household demographics are collected through trained interviewers. This information is then combined with the energy supplier data to estimate the costs and usage for heating, cooling, appliances and other energy uses. The link to this survey results is: http://www.eia.gov/consumption/residential/.

C.3.2   Commercial Building Energy Consumption Survey (CBECS)

The Commercial Buildings Energy Consumption Survey (CBECS) is a national sample survey that collects information on the stock of U.S. commercial buildings, including the energy related characteristics and energy usage data. Commercial buildings involved in this survey are buildings that at least half of the floor space is used for a purpose that is not residential, industrial or agricultural. Similar to RECS, building characteristics and energy usage data are collected from interviews. A follow up Energy Supplier Survey (ESS) is conducted for buildings without accurate data. The link to this survey results is: http://www.eia.gov/consumption/commercial/.

C.3.3   Manufacturing Energy Consumption Survey (MECS)

The Manufacturing Energy Consumption Survey (MECS) is a national sample survey that collects information on the stock of U.S. manufacturing establishment, building characteristics and the energy consumption and expenditures. This is the most recent survey. The link to this survey results is: http://www.eia.gov/consumption/manufacturing/.

There are a lot of information provided as excel files under the "Data" tab on each of the link mentioned above. For RECS, total and average square footage of U.S. homes is available as well as specific category such as Northeast, Midwest, South, West, single-family, multi-family and mobile homes. CBESC provides summary table of total and means of floor space, number of workers, and hours of operation. MECS provides highly energy focused data, such as energy consumption for all purposes by industry and reason, divided into types of fuel such as electricity, residual/distilled oil, natural gas, coal and so on.

C.4   Recommendation

Based on review of materials from NFPA, USFA and USEIA, I recommend a residential occupancy for potential expansion of the assessment tool beyond warehouse buildings. I chose relevance and statistical availability of categories as important criteria to determine the possible expansion of the assessment tool. NFPA and USFA statistics show that one of the most affected occupancy type is residential. Residential occupancy is a leading type not only for the number of fires occurred each year, but also for civilian deaths, injuries, and the property damage. An expansion to include this occupancy type would have a great impact on the scale of fire assessment.

There seems to be more information available on residential buildings. An availability of statistics is important for developing an assessment tool since the model is currently designed to analyze row data. A quick search on USFA's website provided around 30 reports on residential fires, which was 6 times more than the search results on warehouse fires. As mentioned before, a literature review on those reports can be done if necessary. The USEIA website also provides a lot of building information on residential, commercial and manufacturing facilities based on surveys. I believe this would be very helpful as all the data are available freely online as excel files.

Once the exact set of parameters to use in the model is identified, the recommended occupancy type can be further analyzed to see if given parameters can easily be found.

References

1. Commercial Buildings Energy Consumption Survey (CBECS): U.S. Energy Information Administration. https://www.eia.gov/consumption/commercial/ (2015).
2. Fires by Occupancy: Structure, Vehicles, Outside and Other: NFPA research—report and statistics. National Fire Protection Association, 1 Apr 2013. http://www.nfpa.org/research/reports-and-statistics/fires-in-the-us/overall-fire-problem/fires-by-occupancy-structure-vehicles-outside
3. Haynes, H.: Fire loss in the United States. NFPA research—report and statistics. National Fire Protection Association, 1 Sept 2015. http://www.nfpa.org/research/reports-and-statistics/fires-in-the-us/overall-fire-problem/fire-loss-in-the-united-states
4. Manufacturing Energy Consumption Survey (MECS): U.S. Energy Information Administration. http://www.eia.gov/consumption/manufacturing/ (2015)
5. Residential Energy Consumption Survey (RECS): U.S. Energy Information Administration. http://www.eia.gov/consumption/residential/index.cfm (2015)
6. Trends and Patterns of U.S. Fire Losses: NFPA research—report and statistics. National Fire Protection Association, 1 Mar 2015. http://www.nfpa.org/research/reports-and-statistics/fires-in-the-us/overall-fire-problem/trends-and-patterns-of-us-fire-losses
7. U.S. Fire Statistics: Data—publication and library. U.S. Fire Administration. https://www.usfa.fema.gov/data/statistics/ (2015)

Appendix D
Full Enveco Tool Output for Both Case Studies

Case Study 1 Note the the results may differ slightly from Table 9 due to the random nature of the Monte Carlo calculations during updates.

© Fire Protection Research Foundation 2016
F. Amon et al., *Development of an Environmental and Economic Assessment Tool (Enveco Tool) for Fire Events*, SpringerBriefs in Fire, DOI 10.1007/978-1-4939-6559-5

Scenario 1: Fire Service Intervention, Case Study 1.

Fire Service Intervention - No Quantitative Risk Assessment, All data is known

Fire Service Intervention - Economic Assessment

Value	Unit	Description
0	USD	Firefighter fatalities
0	USD	Firefighter injuries
1289642	USD	Property damage
2508828	USD	Job disruption
377310	USD	Direct business interruption
37731	USD	Indirect business interruption
43200	USD	Fire service intervention
33281	USD	Rent reduction
4289992	USD	**TOTAL COST - Fire Service Intervention**

Fire Service Intervention - Environmental Assessment

	Fire emissions	Structure replaced	Contents replaced	Fire department
Global warming (kg CO_2 equiv)	9	1628426	22622	3.3E-03
Acidification (kg SO_2 equiv)	15	6975	30996	2.7E-03
Eutrophication (kg N equiv)	1.8	497	920433	5.4E-03
Ozone depletion (kg CFC-11 equiv)	0	0	93556	1.7E-03
Smog (kg O_3 equiv)	0.4	102885	26199	5.1E-03
Ecotoxicity (CTUe equiv)	0.3	n/a	6409033	3.6E-02
Energy used (MJ)	n/a	19374466	5274720	n/a

Scenario 2: No fire service intervention, Case Study 1.

No Fire Service - Quantitative Risk Assessment
Likelihood for fire spread to adjacent structure

South	North	West	East	Total
0.47	0.00	0.00	0.25	0.72

No Fire Service - Economic Assessment

Value	Unit	Description
0	USD	Firefighter fatalities
0	USD	Firefighter injuries
2219472	USD	Property damage
4317688	USD	Job disruption
649350	USD	Direct business interruption
64935	USD	Indirect business interruption
0	USD	Fire service intervention
57277	USD	Rent reduction
7308721	USD	TOTAL COST - NO Fire Service

No Fire Service - Environmental Assessment

	Fire emissions	Structure replaced	Contents replaced	Fire department
Global warming (kg CO_2 equiv)	15	2802518	38932	n/a
Acidification (kg SO_2 equiv)	26	12003	12003	n/a
Eutrophication (kg N equiv)	3.2	855	1584064	n/a
Ozone depletion (kg CFC-11 equiv)	0	0	161010	n/a
Smog (kg O_3 equiv)	0.7	177065	45089	n/a
Ecotoxicity (CTUe equiv)	0.5	n/a	11029932	n/a
Energy used (MJ)	n/a	33343416	9077782	n/a

Comparison: Scenario 2–Scenario 1, savings due to fire service intervention, Case Study 1.

Savings due to Fire Service Response - QRA

On average: **0.72Warehouses**

Savings - Economic Assessment

ValueUnit		Description
0	USD	Firefighter fatalities
0	USD	Firefighter injuries
929829	USD	Property damage
1808860	USD	Job disruption
272040	USD	Direct business interruption
27204	USD	Indirect business interruption
-43200	USD	Fire service intervention
23996	USD	Rent reduction
3018728	USD	**TOTAL ECONOMIC SAVINGS**

Savings - Environmental Assessment

	Fire emissions	Structure replaced	Contents replaced	Fire department	Total Savings
Global warming (kg CO_2 equiv)	6	1174092	16310	-3.3E-03	**1.2E+06**
Acidification (kg SO_2 equiv)	11	5029	22348	-2.7E-03	**2.7E+04**
Eutrophication (kg N equiv)	1.3	358	663630	-5.4E-03	**6.6E+05**
Ozone depletion (kg CFC-11 equiv)	0	0	67454	-1.7E-03	**6.7E+04**
Smog (kg O_3 equiv)	0.3	74180	18890	-5.1E-03	**9.3E+04**
Ecotoxicity (CTUe equiv)	0.2n/a	n/a	4620899	-3.6E-02	**4.6E+06**
Energy used (MJ)	n/a	13968950	3803062	n/a	**1.8E+07**

Case Study 2 Note the the results may differ slightly from Table 10 due to the random nature of the Monte Carlo calculations during updates.

Scenario 1: Fire Service Intervention, Case Study 2.

Fire Service Intervention - No Quantitative Risk Assessment, All data is known

Fire Service Intervention - Economic Assessment

Value	Unit	Description
0	USD	Firefighter fatalities
0	USD	Firefighter injuries
7719253	USD	Property damage
15016783	USD	Job disruption
1420787	USD	Direct business interruption
142079	USD	Indirect business interruption
1152000	USD	Fire service intervention
199207	USD	Rent reduction
25650108	USD	**TOTAL COST - Fire Service Intervention**

Fire Service Intervention - Environmental Assessment

	Fire emissions	Structure replaced	Contents replaced	Fire department
Global warming (kg CO_2 equiv)	54	9747071	116847	9.0E-03
Acidification (kg SO_2 equiv)	90	41747	164286	7.5E-03
Eutrophication (kg N equiv)	11.0	2974	5360802	1.5E-02
Ozone depletion (kg CFC-11 equiv)	0	0	35064	4.7E-03
Smog (kg O_3 equiv)	2.3	615828	145250	1.4E-02
Ecotoxicity (CTUe equiv)	1.8	n/a	38097627	9.9E-02
Energy used (MJ)	n/a	115967358	4297710	n/a

Scenario 2: No fire service intervention, Case Study 2.

No Fire Service - Quantitative Risk Assessment
Likelihood for fire spread to adjacent structure

South	North	West	East	Total
0.00	0.08	0.00	0.00	0.08

No Fire Service - Economic Assessment

Value	Unit	Description
0	USD	Firefighter fatalities
0	USD	Firefighter injuries
8357571	USD	Property damage
16258547	USD	Job disruption
1538274	USD	Direct business interruption
153827	USD	Indirect business interruption
0	USD	Fire service intervention
215679	USD	Rent reduction
26523899	USD	**TOTAL COST - NO Fire Service**

No Fire Service - Environmental Assessment

	Fire emissions	Structure replaced	Contents replaced	Fire department
Global warming (kg CO_2 equiv)	58	10553073	126510	n/a
Acidification (kg SO_2 equiv)	97	45199	177871	n/a
Eutrophication (kg N equiv)	11.9	3220	5804096	n/a
Ozone depletion (kg CFC-11 equiv)	0	0	37963	n/a
Smog (kg O_3 equiv)	2.5	666752	157261	n/a
Ecotoxicity (CTUe equiv)	1.9	n/a	41247986	n/a
Energy used (MJ)	n/a	125556900	4653095	n/a

Comparison: Scenario 2–Scenario 1, savings due to fire service intervention, Case Study 2.

Savings due to Fire Service Response - QRA

On average: **0.08 Warehouses**

Savings - Economic Assessment

Value	Unit	Description
0	USD	Firefighter fatalities
0	USD	Firefighter injuries
638318	USD	Property damage
1241764	USD	Job disruption
117487	USD	Direct business interruption
11749	USD	Indirect business interruption
-1152000	USD	Fire service intervention
16473	USD	Rent reduction
873791	USD	**TOTAL ECONOMIC SAVINGS**

Savings - Environmental Assessment

	Fire emissions	Structure replaced	Contents replaced	Fire department	Total Savings
Global warming (kg CO_2 equiv)	4	806002	9662	-9.0E-03	8.2E+05
Acidification (kg SO_2 equiv)	7	3452	13585	-7.5E-03	1.7E+04
Eutrophication (kg N equiv)	0.9	246	443294	-1.5E-02	4.4E+05
Ozone depletion (kg CFC-11 equiv)	0	0	2899	-4.7E-03	2.9E+03
Smog (kg O_3 equiv)	0.2	50924	12011	-1.4E-02	6.3E+04
Ecotoxicity (CTUe equiv)	0.1	n/a	3150359	-9.9E-02	3.2E+06
Energy used (MJ)	n/a	9589542	355385	n/a	9.9E+06

References

1. Hall, J.R.: The Total Cost of Fire in the United States. National Fire Protection Association, Quincy (2014)
2. U.S. Fire Statistics. Working for a Fire-Safe America (cited 2016, Available from: https://www.usfa.fema.gov/data/statistics/)
3. Fighting Fire with Facts (cited 2016, Available from: http://www.nfic.org/)
4. World Fire Statistics. Available from: http://www.ctif.org/ctif/world-fire-statistics (2016)
5. Palm, A., et al.: Assessing the environmental fate of chemicals of emerging concern: a case study of the polybrominated diphenyl ethers. Environ. Pollut. **117**, 195–213 (2002)
6. Alaee, M.: Recent progress in understanding of the levels, trends, fate and effects of BFRs in the environmen. Chemosphere **64**, 179–180 (2006)
7. Lönnermark, A., et al.: Emissions from fire-methods, models and measurements. In: Interflam 2007. Royal Holloway College, University of London, Interscience Communications, London (2007)
8. Noiton, D., Fowles, J., Davies, H.: The Ecotoxicity of Fire-Water Runoff. Part II. Analytical Results, p. 23. New Zealand Fire Service Commission (2001)
9. Blomqvist, P., Simonson-McNamee, M.: Large-scale generation and characterisation of fire effluents. In: Stec, A., Hull, R. (eds.) Fire Toxicity, pp. 461–514. Woodhead Publishing Limited, Cambridge (2010)
10. Hamins, A., et al.: Reducing the Risk of Fire in Buildings and Communities: A Strategic Roadmap to Guide and Prioritize Research, p. 171. National Institute of Standards and Technology, Gaithersburg (2012)
11. Standardization, I.O.F.: ISO 14040:2006 Environmental Management—Life Cycle Assessment—Principles and Framework, p. 20. ISO/TC 207/SC 5, Geneva (2006)
12. Standardization, I.O.F.: ISO 14044:2006 Environmental Management—Life Cycle Assessment—Requirements and Guidelines, p. 46. European Committee for Standardization, Geneva (2006)
13. Andersson, P., et al.: Fire-LCA Model: Furniture Study, p. 61. SP Swedish National Testing and Research Institute, Borås (2003)
14. Andersson, P., et al.: Fire-LCA Guidelines—NICe project 04053 (2004)
15. Simonson, M., et al.: Fire-LCA Model: Cables Case Study, p. 129. SP Sveriges Tekniska Forskningsinstitut, Borås (2001)
16. Simonson, M., et al.: Fire-LCA Model: TV Case Study, p. 212. SP, Borås (2000)
17. Hamzi, R., Londiche, H., Bourmada, N.: Fire-LCA model for environmental decision-making. Chem. Eng. Res. Des. **86**(10), 1161–1166 (2008)
18. Chettouh, S., et al.: Interest of the theory of uncertain in the dynamic LCA-fire methodology to assess fire effects. In: 8th International Conference on Material Sciences, CSM8-ISM5 2012. Elsevier, Beirut (2014)

© Fire Protection Research Foundation 2016
F. Amon et al., *Development of an Environmental and Economic Assessment Tool (Enveco Tool) for Fire Events*, SpringerBriefs in Fire, DOI 10.1007/978-1-4939-6559-5

19. Standardization, I.O.F.: ISO/TS 14067:2013 Greenhouse Gases—Carbon Footprint of Products—Requirements and Guidelines for Quantification and Communication. International Organization for Standardization, Geneva (2013)
20. Standardization, I.O.F.: ISO 14046:2014 Environmental Management—Water Footprint— Principles, Requirements and Guidelines, Geneva (2014)
21. Energy, U.D.O.: Commercial Buildings Energy Consumption Survey (CBECS). (cited 2016, The Commercial Buildings Energy Consumption Survey (CBECS) is a national sample survey that collects information on the stock of U.S. commercial buildings, including their energy-related building characteristics and energy usage data (consumption and expenditures). Commercial buildings include all buildings in which at least half of the floorspace is used for a purpose that is not residential, industrial, or agricultural. By this definition, CBECS includes building types that might not traditionally be considered commercial, such as schools, hospitals, correctional institutions, and buildings used for religious worship, in addition to traditional commercial buildings such as stores, restaurants, warehouses, and office buildings). Available from: http://buildingsdatabook.eren.doe.gov/CBECS.aspx (2012)
22. Bare, J.: Tool for the reduction and assessment of chemical and other environmental impacts (TRACI) version 2.1. In: Young, D., Hopton, M. (eds) STD Standard Operating Procedure SOP No. S-10637-OP-1-0, p. 24. US Environmental Protection Agency (2012)
23. User Manual and Transparency Document—Impact Estimator for Buildings, vol. 5. p. 50, Athena Sustainable Materials Institute, Ottawa/Kutztown (2014)
24. What's New in SimaPro 8.1, p. 8. PRé Consultants (2016)
25. Cortez, J.: Environmental Sustainability: Capital Purchasing and Sustainability (2013) (cited 2016, Available from: http://www.iafc.org/onScene/article.cfm?ItemNumber=7146)
26. Franks, A., et al.: Application of QRA in Operational Safety Issues, p. 97. Health and Safety Executive, Stockport (2002)
27. McNamee, M.S., Andersson, P.: Application of a cost–benefit analysis model to the use of flame retardants. Fire Technol **51**(1), 67–83 (2014)
28. Ingason, H., Tuovinen, H., Lönnermark, A.: Industrial fires—a literature survey. In: SP Report 2010:17. SP Technical Research Institute of Sweden, Borås (2010)
29. Heskestad, G.: Luminous heights of turbulent diffusion flames. Fire Saf. J. **5**, 103–108 (1983)
30. Heskestad, G.: A reduced-scale mass fire experiment. Combust. Flame **83**, 293–301 (1991)
31. Mudan, K.S., Croce, P.A.: Fire hazard calculations for large open hydrocarbon fires. In: DiNenno, P.J., et al. (eds.) The SFPE Handbook of Fire Protection Engineering. The National Fire Protection Association (1995)
32. Hendrickson, C., et al.: Economic input-output models for environmental life-cycle assessment. Environ. Sci. Technol. **32**(7), 184A–191A (1998)
33. Matthews, H.S., Small, M.J.: Extending the boundaries of life-cycle assessment through environmental economic input-output models. J. Ind. Ecol. **4**(3), 7–10 (2001)
34. Gritzo, L.A., et al.: The Influence for Risk Factors on Sustainable Development, p. 27. FM Global, Norwood (2009)
35. Fraser-Mitchell, J., Abbe, O., Williams, C.: In: Smith, D. (eds.) An Environmental Impact and Cost Benefit Analysis for Fire Sprinklers in Warehouse Buildings, p. 124. BRE Global, Watford, Herts (2013)
36. Dinaburg, J., Gottuk, D.: Fire Detection in Warehouse Facilities, in Fire Detection in Warehouse Facilities, pp. 1–59. Springer New York (2012)
37. NFPA: NFPA 5000: Building Construction and Safety Code. NFPA, Quincy (2015)
38. NFPA: NFPA 13: Standard for the Installation of Sprinkler Systems. NFPA, Quincy (2015)
39. NFPA: NFPA 80A: Recommended Practice for Protection of Buildings from Exterior Fire Exposures. NFPA, Quincy (2012)
40. Campbell, R.: Structure Fires in US Warehouses. National Fire Protection Association, Quincy (2013)
41. Carlsson, E.: External Fire Spread to Adjoining Buildings—A Review of Fire Safety Design Guidance and Related Research. Lund University, Lund (1999)

42. Thomas, G.C., et al.: Post-earthquake fire spread between buildings—estimating and costing extent in Wellington. In: Evans, D.D. (ed) Fire Safety Science—Proceedings of the 7th International Symposium, pp. 691–702. International Association for Fire Safety Science, Worcester (2002)
43. Wickström, U.: Heat Transfer in Fire Technology. Luleå Technical University (2012)
44. Lönnermark, A., Ingason, H. (2010) Fire Spread Between Industry Premises. SP Report 2010:18. SP Technical Research Institute of Sweden, Borås (2010)
45. Boardman, A., et al.: In: Battista, D. (ed.) Cost-Benefit Analysis: Concepts and Practice. Prentice Hall, Upper Saddle River (2011)
46. European Commission: ILCD Handbook—General Guide for Life Cycle Assessment—Detailed Guidance, p. 417. Joint Research Centre—Institute for Environment and Sustainability (2010)
47. Blomqvist, P., Rosell, L., Simonson, M.: Emissions from fires Part II: simulated room fires. Fire Technol **40**(1), 59–73 (2004)
48. Blomqvist, P., Lönnermark, A., Simonson, M.: Miljöbelastning vid bränder och andra olyckor- Utvärdering av provtagning och analyser. In: The Environmental Impact of Fires and Other Accidents. An Evaluation of Samples and Analyses, p. 52. Räddningsverket, Borås (2004)
49. Bare, J.: In: N.R.M.R. Laboratory (ed.) Tool for the Reduction and Assessment of Chemical and Other Environmental Impacts (TRACI): User's Guide and System Documentation, p. 79. U.S. Environmental Protection Agency, Cincinnati (2003)
50. Huijbregts, M., et al.: USEtox User Manual, p. 23 (2010)
51. Duval, R.: Storage Warehouse, Phoenix, AZ. National Fire protection Association, Quincy (2002), 2 Aug 2000
52. NFPA: Fire Investigations: Storage Warehouse, Phoenix, AZ. National Fire Protection Association, Quincy (2002), 2 Aug 2000
53. Duval, R.: Large Loss Fire in Sprinklered Warehouse. In: SFPE—Great Atlanta Chapter Annual Conference (2016)
54. Ott, H.: Louisville Structure Fire Investigation Report. Incident number 15-80-000597 (2015)

Printed in the United States
By Bookmasters